CLASSICAL MECHANICS

Richard A. Matzner

Center for Relativity
The University of Texas at Austin

Lawrence C. Shepley

Center for Relativity
The University of Texas at Austin

PRENTICE HALL, *Englewood Cliffs, New Jersey 07632*

Library of Congress Cataloging-in-Publication Data

MATZNER, RICHARD A. (RICHARD ALFRED). date
    Classical mechanics / Richard A. Matzner, Lawrence C. Shepley. p.   cm.
    Includes index.
    ISBN 0-13-137076-6
    1. Mechanics.   2. Physics.   3. Astrophysics.   4. Mathematical physics.   I. Shepley,
    Lawrence C., date   II. Title.
    QC125.2.M37   1991
    531—dc20   90-20798
                                                                            CIP

Editorial/production supervision and interior design: Virginia Huebner
Cover design: Lundgren Graphics, Ltd.
Prepress buyer: Paula Massenaro
Manufacturing buyer: Lori Bulwin
Cover photo: Courtesy of NASA

 © 1991 by Prentice-Hall, Inc.
A Division of Simon & Schuster
Englewood Cliffs, New Jersey 07632

Printed in the United States of America
10 9 8 7 6 5 4 3 2 1

ISBN 0-13-137076-6

Prentice-Hall International (UK) Limited, *London*
Prentice-Hall of Australia Pty. Limited, *Sydney*
Prentice-Hall Canada Inc., *Toronto*
Prentice-Hall Hispanoamericana, S.A., *Mexico*
Prentice-Hall of India Private Limited, *New Delhi*
Prentice-Hall of Japan, Inc., *Tokyo*
Simon & Schuster Asia Pte. Ltd., *Singapore*
Editora Prentice-Hall do Brasil, Ltda., *Rio de Janeiro*

To   SARA

To   LONA and JORAM

CONTENTS

Preface

Chapter **0**   **Paradigm**   1

Chapter **1**   **Mathematical and Kinematical Preliminaries**   3

    1.1   Manifolds and Vectors, *4*
    1.2   Gradients and 1-Forms, *11*
    1.3   Tensors, *14*
    1.4   Configuration Space and Kinetic Energy, *17*
    1.5   Holonomic Constraints, *19*
    1.6   Examples, *24*
    Exercises, 25

Chapter **2**   **Spacetime**   32
    Exercises, 36

Chapter **3**   **Lagrangians and Lagrangian Systems**   38
    Exercises, 43

Chapter **4**   **Central Force Fields**   48

    4.1   General Considerations, *49*
    4.2   The Two-Body Problem, *51*
    4.3   The Orbit Equation, *54*
    4.4   The Coulomb Force, *56*
    4.5   Classical Scattering Theory, *58*
    4.6   The Three-Body Problem, *64*
    Exercises, 70

Chapter **5**   **Equivalent Lagrangians**   75

5.1   The Free Particle, *76*

5.2   One-Dimensional Case, *79*

5.3   Many Dimensions, *82*
      Exercises, 84

Chapter **6**   **Rotations and Spinors**   86
      Exercises, 97

Chapter **7**   **Dynamics of Rigid Body Motion**   99

7.1   Equations of Motion in the Body Frame, *104*

7.2   Particle Dynamics in Rotating Frames, *108*

7.3   Rotating Frames and Larmor's Theorem, *111*
      Exercises, 113

Chapter **8**   **Hamiltonian Systems**   119

8.1   Hamilton's Equations, *119*

8.2   Weiss Action Principle and Noether's Theorem, *124*

8.3   Phase Space and Phase Spacetime, *127*

8.4   Examples, *134*
      Exercises, 147

Chapter **9**   **Poisson Brackets**   146

9.1   Poisson Brackets and the Equations of Motion, *146*

9.2   Forms and Integration, *150*
      Exercises, 162

Chapter **10**   **Canonical Transformations**   163
      Exercises, 170

Chapter **11**   **Infinitesimal Canonical Transformations**   173

11.1   The Hamilton–Jacobi Theory, *173*

11.2   Hamilton–Jacobi Theory—Time–Independent Case, *181*

11.3   Hamilton–Jacobi Theory and Wave Mechanics, *183*
         Exercises, 185

Chapter **12**   **Separable Hamiltonians and the Action–Angle Formulation**   188

12.1   Introduction, *188*

12.2   Proof of the Adiabatic Invariance of Action Variables, *199*

12.3   Example: The Kepler Problem as an Action–Angle Problem, *201*
         Exercises, 210

Chapter **13**   **Canonical Perturbation Theory**   214

13.1   Canonical Perturbations: An Example, *221*
         Exercises, 225

Chapter **14**   **Small Oscillations and Continuous Systems**   227

14.1   Example: Coupled Harmonic Oscillators, *230*

14.2   Small Oscillations of Many-Body and of Continuous Systems;
         Classical Fields, *232*

14.3   Periodic Boundary Conditions, *238*
         Exercises, 239

Index   243

PREFACE

Physics is the study of the fundamental laws of nature, but what constitutes a law and which laws are taken to be fundamental are matters of evolving consensus among physicists. Almost all agree that the subject called classical mechanics forms an important part of graduate education in physics. It is also a most enjoyable part, for it brings together practical applications and subtle mathematical techniques.

For many years we have been helping graduate students learn theoretical physics. Our own field is general relativity, a field which requires the techniques of differential geometry. We have found that the same techniques, in a somewhat informal style, are not only an appealing way to treat classical mechanics but are also a good way to teach the subject.

The students for whom this book is designed are first year graduate students in physics. We presume that as undergraduates they have had a course at the level of Ralph Baierlein's excellent text **Newtonian Dynamics** (McGraw-Hill, 1983) and are familiar with the usual undergraduate courses in quantum mechanics and electrodynamics. We recognize that these students may not have taken any graduate level physics or mathematics, and so we have been explicit in developing our ideas. At the same time, we recognize that a book which is too long can be discouraging. Our text is meant to be covered in one semester, but we have given somewhat more than the number of exercises we usually assign.

We thank many: Our colleagues, students, teachers have given good and helpful advice. Our reviewers and editors have been critical; we appreciate that. Our friends, relatives, pets have been most supportive. All are too many to mention explicitly, but we would like to acknowledge and thank our reviewers: Ralph Baierlein (Wesleyan University, Middletown, CT), Sumner P. Davis (University of California, Berkeley), Charles J. Goebel (University of Wisconsin, Madison), David R. Harrington (Rutgers University, Piscataway, NJ), James C. Ho (Wichita State University), Lawrence S. Pinsky (University of Houston), Mitchell J. Sweig (Northeastern Illinois University, Chicago), Gary Williams (University of California, Los Angeles), and Lowell Wood (University of Houston). None of course bear any blame for our shortcomings.

Readers: Please tell us your suggestions, corrections, and comments. We Texans are friendly and will be most grateful for them.

Austin, Texas

Richard A. Matzner
Lawrence C. Shepley

0

PARADIGM

This is a graduate text in classical mechanics. Few subjects in physics successfully embrace as wide a range of applications as does classical mechanics. Classical mechanics is known to be a limit of quantum mechanics and in many contexts provides answers that have extremely high fractional accuracy, since human-sized systems typically have an action very much longer than the quantum unit, $h = 6.63 \times 10^{-27}$ erg s. Even for atomic or smaller systems, classical mechanics provides a valuable approximation tool, giving an estimate of expected behavior and a feel for magnitudes and units.

Classical mechanics was the driving force behind the flowering of 18th and 19th century analysis, and many features of classical mechanics are pursued in the new revolution of computational physics. Because it is a limit of quantum mechanics, there has been a flow of innovation and insight from wave mechanics back to classical mechanics since the 1930's.

The standard syllabus of classical mechanics is covered here, but it is extended to topics that the student will find helpful in future encounters with modern aspects of the subject. We have tried to balance direct physical reasoning

1

and powerful mathematics. Mathematics is included because it is powerful, and we hope we avoided making it impenetrable. All problems in physics demand some subtlety of intuition. The formal structures we introduce reinforce that intuition, and in most chapters, we include problems for the students to exercise it.

1

MATHEMATICAL AND KINEMATICAL PRELIMINARIES

The mathematical description of classical systems has proceeded from the Aristotelian view through the remarkable insights of Newton and his contemporaries into the great flowering of the eighteenth- and nineteenth-century mathematical dynamics of Hamilton, Lagrange, and Jacobi. It is still an active research field today and has application in celestial mechanics, statistical mechanics, and the foundations of quantum mechanics and quantum field theory. Newton's equations are treated in introductory and upper-level undergraduate texts. In a graduate text, the mathematical techniques are described in much more detail, as are applications to very general and even esoteric systems. Nonetheless, this is a physics text, designed for a graduate course on the mathematical and physical concepts of mechanics and their relation to other branches of physics.

This first chapter sets notation and serves as a partial review of material covered in undergraduate mathematics and physics courses. Our definitions and our proofs are not meant to be rigorous in a strict mathematical sense, but they are given in a precise way so as to allow their use in very general physical problems. The first three sections of the chapter are an introduction to mathematical concepts, taken from differential geometry, which will be used throughout the book and extended in future chapters. The other sections deal with the configuration spaces of general physical systems, defined in the language of manifolds.

1.1   Manifolds and Vectors

Classical physics assumes that time and space form a continuous, differentiable arena for physics. In the Galilean view space and time are differentiable manifolds (of dimensions 3 and 1, respectively); in the relativistic view, spacetime is a differentiable manifold of dimension 4. In the mechanics of a many-particle system, it is appropriate to use a manifold of many dimensions to describe the instantaneous state of the system.

A manifold is a space that looks "locally" like a Euclidean space of finite dimension. This means that at any particular point of a manifold, we can if we wish erect a "rectangular" coordinate sytem $x, y, z, \ldots, u, v, w$ of some dimension, and if we consider small enough displacement from our chosen point, nearby points in the manifold can be coordinatized using this coordinate system. The word "rectangular" is put in quotes because we don't use any concept of right angles or of length; we simply are uniquely labeling points of the manifold by values of $x, y, z, \ldots, u, v, w$.

A space $\mathcal{M}$ is a **manifold** if each point of $\mathcal{M}$ lies within at least one neighborhood $\mathcal{N}$, the points of which can be put into a 1–1 correspondence with points of an open set of $\mathbf{R}^n$ (the space described by n real-valued coordinates). This correspondence defines a set of functions $\{\phi^\alpha\}$ on $\mathcal{N}$, with $\alpha = 1, \ldots, n$, where n is the dimension of the Euclidean space $\mathbf{R}^n$. The values taken by the ϕ^α at the point P in $\mathcal{N}$ (which give the point in $\mathbf{R}^n$, which is the image of P) are called the **coordinates** of the point P. $\mathcal{N}$ is called a **coordinate patch**. Notice that we have said nothing about how the coordinates at one point are related to those at nearby points. Essentially, we take our first point and erect a coordinate system there. A sufficiently nearby point will lie within this first coordinate "patch." A new coordinate patch can be erected at this second point, and the two patches will overlap. In the overlap there will be points that have coordinates from each of the two coordinate patches. We can call the two patches $(\mathcal{N}, \phi^\alpha)$ and $(\bar{\mathcal{N}}, \bar{\phi}^\beta)$, where $\mathcal{N}$ and $\bar{\mathcal{N}}$ are the names of the neighborhoods of each of the points, and $\phi^1, \phi^2, \ldots, \phi^N$ (abbreviated ϕ^α, $\alpha = 1, \ldots, N$) and $\bar{\phi}^1, \bar{\phi}^2, \ldots, \bar{\phi}^N$ (i.e., $\bar{\phi}^\beta$, $\beta = 1, \ldots, N$) are the labels we use to stand for the coordinates in each of these two patches. In the overlap, each point has coordinates ϕ^α (a specific and unique set of numbers for each point), as well as coordinates $\bar{\phi}^\beta$ (a specific set of numbers different for each point but not necessarily different from some set ϕ^α). Hence we have a coordinate transformation: $\phi^\alpha = \phi^\alpha(\bar{\phi}^\beta)$; $\bar{\phi}^\beta = \bar{\phi}^\beta(\phi^\alpha)$. Each ϕ^α for a particular point is a function of (i.e., is uniquely determined by) all the $\bar{\phi}^\beta$ for that point, and vice versa. Figure 1.1 gives an example of a manifold. For the purposes of classical physics, a certain degree of smoothness is always assumed for the structures under discussion. Smoothness, in particular differentiability, of the space is assumed in all cases.

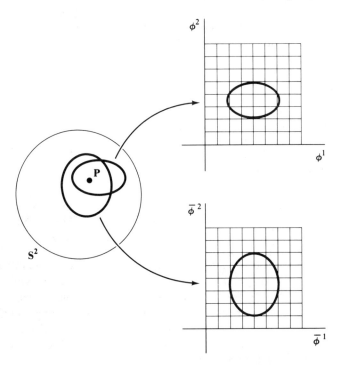

Figure 1.1. The manifold $\mathbf{S}^2$, the two-dimensional sphere. The point **P** is shown as being in two overlapping coordinate neighborhoods.

$\mathcal{M}$ is a **differentiable manifold** if there is at least one covering of $\mathcal{M}$ by coordinate patches such that in each overlap of the patches the coordinates are mutually related by differentiable functions. Such a covering is called a differentiable **atlas**. The concept of differentiable manifold may be expressed in terms of **coordinate transformations**. Let $(\mathcal{N}, \phi^\alpha)$ and $(\bar{\mathcal{N}}, \bar{\phi}^\beta)$ be two coordinate patches in the differentiable atlas. Then in $\mathcal{N} \cap \bar{\mathcal{N}}$, the ϕ^α are differentiable functions of the $\bar{\phi}^\beta$, and vice versa. [If the atlas is such that these functions are m times differentiable (C^m), then $\mathcal{M}$ is said to be a C^m-manifold. We assume that m is large enough so that differentiability problems will not matter in our physical discussion.] The coordinate transformation $\{\phi^\alpha\} \rightarrow \{\bar{\phi}^\beta\}$ is only defined in $\mathcal{N} \cap \bar{\mathcal{N}}$, but in general we will follow the informal practice of not stating explicitly either $\mathcal{N}$ or $\bar{\mathcal{N}}$. New coordinates $\bar{\bar{\phi}}^\gamma$ may be defined by taking n independent, differentiable functions of any of the coordinate functions in the atlas; the neighborhood $\bar{\bar{\mathcal{N}}}$ on which these functions may be taken as coordinates is any neighborhood on which they are truly independent and differentiable.

Here is an example of a coordinate transformation. First, ordinary Euclidean space is a manifold. Hence the rectangular coordinates x, y, z around one origin and the $\bar{x}, \bar{y}, \bar{z}$ around another are related to one another by a differentiable coordinate transformation. The spherical coordinate set r, θ, ϕ is related to x, y, z by

$$r = (x^2 + y^2 + z^2)^{1/2},$$
$$\theta = \arccos[z(x^2 + y^2 + z^2)^{-1/2}],$$
$$\phi = \arctan \frac{y}{x},$$

which defines each of r, θ, ϕ as a function of the x, y, z; the inverse transformation is

$$x = r \sin\theta \cos\phi,$$
$$y = r \sin\theta \sin\phi,$$
$$z = r \cos\theta.$$

Notice that this coordinate transformation is not differentiable at the point $r = 0$. For $r = 0$, different values of θ, ϕ map to the same point, $x = y = z = 0$.

The most fundamental concept in mechanics is that of the **path** of a structureless particle (a point particle) representing the state of a mechanical system. This may be literally a point particle, or the particle may represent some abstraction of a physical system. We will often use this abstraction idea in the sequel.

The particle can take a position in some space; the possible positions it can take are represented by the points of a manifold. A possible path of the particle is then given by describing which position the particle occupies as a function of some parameter, say the time. A possible path is a continuous (in fact, at least twice differentiable) **curve** $\mathcal{C}$, $\mathcal{C}$ being a map from some interval in the real line (the **parameter space**) into the manifold $\mathcal{M}$, which is the arena of the dynamics.

In the Newtonian mechanics of a single, structureless object, the parameter is the universal time and the dynamical space is three-dimensional Euclidean space. In relativity, the dynamics of a single object take place in a four-dimensional spacetime, and the parameter is typically chosen to be the time shown by some internal clock moving along the curve, the **proper time** of an observer moving along with the object. In the coordinate system $\{\phi^\alpha\}$, $\mathcal{C}$ is described by n functions $\phi^\alpha(t)$, t being the path parameter.

To proceed with the development of elementary mechanics, we next need a concept of **velocity**. Velocity, of course, is a **vector**; that is, it is an infinitesimal displacement (divided in this case by an associated infinitesimal time). Some of the development we will give below depends on the additional property of vectors, namely, every vector is the tangent to some parameterized curve, and in fact each vector at a point can be defined by giving all the curves through that point to which it is a vector. The velocity vector is the **tangent vector** at a particular point on $\mathcal{C}$, that is, at a particular parameter value t.

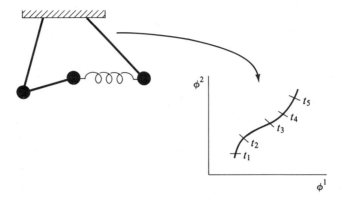

Figure 1.2. An object, in this case several mass points connected by rods and springs, is described by n coordinates, and the changes in its configuration are described by a path in an n-dimensional manifold with path parameter t.

We will view the tangent vector as a **differential operator** acting on an arbitrary test function. In these terms, the tangent vector $\mathbf{V}$ to the curve $\mathcal{C}$ tells what effect an infinitesimal displacement along $\mathcal{C}$ has on the value of an arbitrary test function $\theta = \theta(\mathcal{P})$. When $\phi^\alpha(t)$ are the coordinates in a neighborhood N of the point parameterized by t, $\mathbf{V}(\theta)$ is a real number whose value is $\mathbf{V}(\theta) = d\theta\big(\phi^\alpha(t)\big)/dt$. Thus $\mathbf{V}$ is a differential operator, namely,

$$\mathbf{V} = \frac{d}{dt}, \tag{1.1}$$

and $\mathbf{V}$ gives the rate of change along $\mathcal{C}$ of arbitrary test functions. Note that we use boldface to denote a vector. This operator d/dt may be expressed in terms of the partial differentiation operators $\partial/\partial\phi^\alpha$ within a given coordinate system. The result, which makes use of the chain rule of differentiation, defines n functions V^α, the **components** of $\mathbf{V}$ in the coordinate system:

$$\mathbf{V} = \frac{d}{dt} = \sum_{\alpha=1}^{n} V^\alpha \frac{\partial}{\partial\phi^\alpha}, \tag{1.2}$$

where

$$V^\alpha = \frac{d\phi^\alpha}{dt}. \tag{1.3}$$

Clearly, the values of V^α depend on the coordinate system chosen, even though $\mathbf{V}$ is a geometrical object independent of coordinates, as (1.1) shows. We will follow the customary, though incorrect, usage that calls V^α the vector (even though the V^α simply represent the components of $\mathbf{V}$ in a coordinate system). When the geometrical nature of a vector must be emphasized, we will make a distinction between the vector and its components.

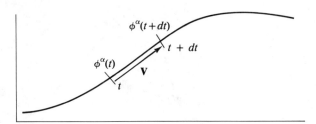

Figure 1.3. The tangent vector $\mathbf{V}$ compares the value of a function f at t with its value at $t+dt$. Hence $\mathbf{V} = \dfrac{d}{dt}$. The value of f at t is really $f(\phi(t))$, and the value of f at $t+dt$ is really $f(\phi(t+dt))$. Hence $\mathbf{V}$ can be expressed in terms of its components as $\mathbf{V} = \dfrac{d}{dt} = \mathbf{V}^{\alpha}\dfrac{\partial}{\partial\phi^{\alpha}}$. $\mathbf{V}$ is best pictured as an arrow.

At this point we explicitly introduce two conventions:

RANGE CONVENTION

An index $\alpha, \beta, \ldots$, takes on the values $\alpha, \beta, \ldots = 1, \ldots, n$, where n is fixed. This convention means that an equation such as

$$A^{\alpha} + B^{\alpha} = C^{\alpha} \tag{1.4}$$

stands for n ordinary equations. We will use both superscripts and subscripts. In an algebraic expression involving two or more terms, indices that appear once in each term are called **free indices**. In Eq. (1.4), α is the free index. The free indices in the various terms must agree as to both name and position.

A term may also have indices that appear in pairs, one a superscript and the other a subscript. Such indices are called **dummy indices**. Dummy indices appear in the second convention we will use:

SUMMATION CONVENTION

A pair of dummy indices implies a summation over the range of the indices:

$$A^{\alpha}B_{\alpha} \equiv \sum_{\alpha=1}^{n} A^{\alpha}B_{\alpha}, \tag{1.5}$$

$$C^{\alpha}{}_{\alpha} \equiv \sum_{\alpha=1}^{n} C^{\alpha}{}_{\alpha}. \tag{1.6}$$

This convention is a very efficient one. The summation index name is clearly arbitrary:

$$A^{\alpha}B_{\alpha} = \sum_{\alpha=1}^{n} A^{\alpha}B_{\alpha} = \sum_{\beta=1}^{n} A^{\beta}B_{\beta} = A^{\beta}B_{\beta}, \tag{1.7}$$

hence the name *dummy index*.

An expression such as $R^{\alpha}{}_{\alpha\beta}$ is too confusing to be allowable, and hence triples of indices are not allowed. In addition, we will see that pairs of indices (implying a sum), both of which are superscripts or subscripts, do not make sense geometrically and thus are also not allowed. For example,

$$A^{\alpha} + B^{\beta\alpha}C_{\beta} = D^{\alpha}$$

is a legitimate expression. So is $A^{\alpha} + B^{\alpha\beta}C_{\beta}$, even though the specific values may be different from the first expression. Each expression stands for n expressions (for $\alpha = 1, \ldots, n$). An expression such as $A^{\alpha} + B_{\alpha\beta}C^{\beta}$ is not correct.

No convention is universally useful. When necessary for clarity, we will drop the convention explicitly in the text or use (NO SUM) in an expression. Thus $T^{\alpha}{}_{\alpha}$(NO SUM) stands for the n expressions $T^{1}{}_{1}, T^{2}{}_{2}, \ldots$, whereas $T^{\alpha}{}_{\alpha}$ stands for the single term $\sum_{\alpha=1}^{n} T^{\alpha}{}_{\alpha}$. In addition, we sometimes will use other types of indices such as $i, A, \ldots$, if different ranges are indicated. One common convention, for example, is to use Latin indices $i, j, \ldots$, having the range $1, 2, 3$, when dealing with ordinary three-dimensional space.

As defined above, all vectors located at a given point P form a finite-dimensional linear space, the **tangent space** at P, T_P. The dimension n of T_P is the same as that of $\mathbf{R}^n$ used in the definition of the manifold. A complete set of n linearly independent vectors may be chosen as a basis for this vector space. That some set of vectors at a point P, say the set $\{\mathbf{e}_{(\alpha)}(P), \ \alpha = 1, \ldots, n\}$, forms a basis means that any vector in the tangent space at that point can be expressed as a finite linear combination of such basis vectors. [In usual Euclidean coordinates, one often defines basis vectors, $\hat{\mathbf{i}}, \hat{\mathbf{j}}, \hat{\mathbf{k}}$, and writes

$$\mathbf{B} = B^1\hat{\mathbf{i}} + B^2\hat{\mathbf{j}} + B^3\hat{\mathbf{k}},$$

with $B^i, i = 1, 2, 3$, the components of $\mathbf{B}$ in the basis $\{\hat{\mathbf{i}}, \hat{\mathbf{j}}, \hat{\mathbf{k}}\}$.] In general, any vector $\mathbf{B}$ in the tangent space may be expressed as a linear combination of the basis vectors $\mathbf{B} = B^{\alpha}\mathbf{e}_{(\alpha)}$; the B^{α} are the components of $\mathbf{B}$ in this basis. Notice that this form uses the summation convention.

This prescription of the tangent space and the basis may be repeated in a continuous way at every point in M. M and the tangent space at each point form the **tangent bundle** over M. The basis vectors form **vector fields** on M. We in general denote the set of basis fields $\{\mathbf{e}_{(\alpha)}\}$.

The coordinate functions ϕ^{α} give a possible basis field set. We define the basis vectors to be $\partial/\partial\phi^{\alpha}$. Here the notation "$\partial$" means *partial* differentiation, but otherwise we are using the operator properties of Eq. (1.1). In other words, $\partial/\partial\phi^{\alpha}$ is the vector that is tangent to the curve $\mathbf{\Phi}^{\alpha}$ defined as the curve along which ϕ^{α} increases but all the other coordinates remain unchanged. To define a tangent vector, one must parameterize the curve; $\partial/\partial\phi^{\alpha}$ is the tangent obtained when the curve $\mathbf{\Phi}^{\alpha}$ is parameterized by ϕ^{α}.

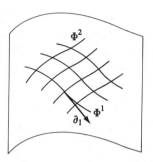

Figure 1.4. The curve Φ^1 is the curve along which all the other coordinates are constant. It is parameterized by ϕ^1, and thus the tangent vector is

$$\boldsymbol{\partial}_1 = \frac{\partial}{\partial \phi^1}.$$

The rectangular coordinate basis $\hat{\mathbf{i}}, \hat{\mathbf{j}}, \hat{\mathbf{k}}$ referred to earlier is an example of such a basis:

$$\hat{\mathbf{i}} = \frac{\partial}{\partial x},$$

$$\hat{\mathbf{j}} = \frac{\partial}{\partial y},$$

$$\hat{\mathbf{k}} = \frac{\partial}{\partial z}.$$

For instance, a vector $\mathbf{A}$ with components 5 units in the x-direction (i.e., the $\hat{\mathbf{i}}$-direction) is

$$\mathbf{A} = 5\hat{\mathbf{i}} = 5\frac{\partial}{\partial x}.$$

To see that this is a correct representation of this vector, ask how it changes a typical test function, the function $\theta = a + bx + cy$ for instance. This yields

$$\mathbf{A}(\theta) = 5b,$$

which correctly gives the change in θ when the x-coordinate increases by 5.

If, instead, we let $\mathbf{A}$ act on a more complicated function, say, $\psi = ex^3 + z$, the result is

$$\mathbf{A}(\psi) = 15ex^2.$$

The value of the result clearly depends on the location at which the evaluation takes place; this example also shows the local derivative nature of $\mathbf{A}$. It gives the change in ψ, over a range (in this case) of 5, assuming that ψ continues to change as its local behavior indicates.

When no confusion arises, we will use the simplified notation ∂_α for $\partial/\partial\phi^\alpha$. Because of the invertibility of the map defined by the collection $\{\phi^\alpha\}$, the set $\{\partial_\alpha\}$ is complete and independent. Such a basis is called a **coordinate basis**. A general basis can be expressed as a set of linear combinations of elements of a coordinate basis:

$$\mathbf{e}_{(\alpha)} = A_\alpha{}^\beta \partial_\beta, \tag{1.8}$$

where $A_\alpha{}^\beta$ is a nonsingular square matrix function of position, and again we note that the summation convention implies a summation in (1.8) over the dummy index β; α is a free index indicating which vector is meant in (1.8).

A final, cautionary word: ∂_α is a vector. Its components are denoted by using superscripts. Thus for vector number 3, ∂_3, the components are the n functions $\partial_3{}^\alpha$. In this case, the components are simple, for $\partial_3{}^\alpha$ has the value 1 when $\alpha = 3$ and zero for other values of α. In fact, if we denote the β component of the αth vector by $\partial_\alpha{}^\beta$, we have

$$\partial_\alpha{}^\beta = \delta_\alpha{}^\beta \equiv 1 \text{ when } \alpha = \beta \text{ and } 0 \text{ otherwise.} \tag{1.9}$$

The expression $\delta_\alpha{}^\beta$ is called the **Kronecker delta**.

1.2   Gradients and 1-Forms

Before defining the physical concepts of kinetic and potential energy, we introduce a few more of the mathematical tools we will use. First, the gradient. Associated with a differentiable function f on M are the contour surfaces of f, the set of $(n-1)$-dimensional surfaces on which f is constant. The change of f between adjacent surfaces leads to one way of defining the gradient of f. However, we shall see that there are definite advantages to defining $\mathbf{d}f$, the gradient of f, as an operator on vectors.

The definition is based on how a vector acts on functions. A vector $\mathbf{A}$ defines a displacement along the curve to which $\mathbf{A}$ is locally tangent. As in (1.2), we evaluate the numerical result of $\mathbf{A}$, considered as an operator on the function f:

$$\mathbf{A}(f) = A^\alpha \frac{\partial f}{\partial \phi^\alpha}. \tag{1.10}$$

This value of this operation is proportional to the difference in the values that f takes at the point with the coordinate values $\phi^\alpha + \varepsilon A^\alpha$ and at the point P with coordinates ϕ^α:

$$\mathbf{A}(f) = \lim_{\varepsilon \to 0} \frac{1}{\varepsilon} \left[f(\phi^\alpha + \varepsilon A^\alpha) - f(\phi^\alpha) \right]. \tag{1.11}$$

The result $\mathbf{A}(f)$ is linear in f; it is also linear in $\mathbf{A}$. We can just as well view (1.10) as defining a linear operator $\mathbf{d}f$ that acts on vectors:

$$\mathbf{d}f(\mathbf{A}) \equiv \mathbf{A}(f). \tag{1.12}$$

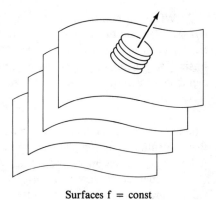

Surfaces f = const

Figure 1.5. df is best represented as a stack of small surfaces parallel to the contour surfaces f = const. Often this stack is represented by an arrow pointing through it.

$\mathbf{d}f$ is a linear function on the set of vectors. It is called the gradient of f. $\mathbf{d}f$ itself is called a differential form, or a **1-form** (higher-ranked forms will be introduced below). The set of all such linear functions or 1-forms constitutes a vector space. A basis set of 1-forms is given by $\{\mathbf{d}\phi^\alpha\}$, the gradients of the coordinate functions ϕ^α. This basis will be called a coordinate basis. Notice that

$$\mathbf{d}\phi^\alpha \left(\frac{\partial}{\partial \phi_\beta} \right) = \frac{\partial \phi^\alpha}{\partial \phi_\beta} = \delta_\beta{}^\alpha, \tag{1.13}$$

where $\delta_\beta{}^\alpha$ is the Kronecker delta introduced above.

Expressed in terms of this basis, our gradient $\mathbf{d}f$ is expressible as:

$$\mathbf{d}f = (df)_\beta \, \mathbf{d}\phi^\beta. \tag{1.14}$$

This is analogous to the corresponding expansion for a vector [cf. Eq. (1.2)] in the coordinate basis $\{\partial/\partial\phi^\alpha\}$:

$$\mathbf{A} = A^\alpha \frac{\partial}{\partial \phi^\alpha}. \tag{1.15}$$

The coefficients $(\mathbf{d}f)_\beta$ in Eq. (1.14) and the coefficients A^α in (1.15) are simply functions; they are not operators. In particular, $\mathbf{d}\phi^\beta$ does not act on A^α, only on $\partial/\partial\phi^\alpha$.

Then, from Eqs. (1.12) and (1.10),

$$\mathbf{d}f(\mathbf{A}) \equiv A(f) = A^\alpha \frac{\partial f}{\partial \phi^\alpha}, \tag{1.16}$$

while from Eqs. (1.14) and (1.15),

$$\mathbf{d}f(\mathbf{A}) = (df)_\beta \mathbf{d}\phi^\beta \left(A^\alpha \frac{\partial}{\partial \phi^\alpha} \right)$$

$$= (df)_\beta A^\alpha \mathbf{d}\phi^\beta \left(\frac{\partial}{\partial \phi^\alpha} \right) \qquad (1.17)$$

$$= (df)_\beta A^\alpha \delta_\alpha{}^\beta$$

$$= (df)_\alpha A^\alpha.$$

Comparing (1.16) with (1.17) shows that the components $(df)_\beta$ of $\mathbf{d}f$, in the coordinate basis, are

$$(df)_\beta = \frac{\partial f}{\partial \phi^\beta}.$$

Whenever a basis set of 1-forms has the property of yielding a δ-symbol when acting on a set of basis vectors, we say that the two sets are **dual** to one another. The set dual to the vector basis $\{\mathbf{e}_{(\alpha)}\}$, defined in general in Eq. (1.8), will be denoted $\{\boldsymbol{\omega}^\beta\}$. This set $\{\boldsymbol{\omega}^\alpha\}$ can be expressed as a linear combination of the coordinate 1-forms $\mathbf{d}\phi^\sigma$:

$$\boldsymbol{\omega}^\beta = B^\beta{}_\sigma \, \mathbf{d}\phi^\sigma. \qquad (1.18)$$

Because the $\{\boldsymbol{\omega}^\beta\}$ are dual to the $\{\mathbf{e}_{(\alpha)}\}$, we have by definition $\boldsymbol{\omega}^\beta(\mathbf{e}_{(\alpha)}) = \delta_\alpha{}^\beta$, which means that

$$B^\beta{}_\sigma A_\alpha{}^\sigma = \delta_\alpha{}^\beta, \qquad (1.19)$$

where $A_\alpha{}^\sigma$ is defined in (1.8).

The definition we just gave of $\mathbf{d}f$ is easily related to the notion that $\mathbf{d}f$ is a small change in f. Suppose that we define a point P by its coordinates ϕ^α. Now let ϕ^α be changed by an amount $\delta\phi^\alpha$. Then f changes by

$$\delta f = \frac{\partial f}{\partial \phi^\alpha} \delta\phi^\alpha. \qquad (1.20)$$

The operator $\mathbf{d}f$ is a **differential form**. The names **covariant vector** or **covector** are also used, but we mainly stick to the term form to emphasize the difference between $\mathbf{d}f$ and a vector $\mathbf{A}$ (which is often called a **contravariant vector**).

A general 1-form is a linear combination of gradients. Thus

$$\boldsymbol{\beta} = \beta_\sigma \, \mathbf{d}\phi^\sigma \qquad (1.21)$$

in the basis $\{\mathbf{d}\phi^\sigma\}$. We use subscripts for the components of 1-forms β_σ to distinguish them from vector components. Be careful to distinguish 1-form component indices from indices used in a basis $\{\mathbf{e}_{(\alpha)}\}$. As with vectors, we customarily call the components, here β_σ, the 1-form.

We will often denote partial differentiation with a comma:

$$f_{,\alpha} \equiv \frac{\partial f}{\partial \phi^\alpha} = \partial_\alpha f. \tag{1.22}$$

A second derivative is

$$f_{,\alpha\beta} \equiv \frac{\partial^2 f}{\partial \phi^\alpha \, \partial \phi^\beta}. \tag{1.23}$$

For functions with continuous second derivatives, we have

$$f_{,\alpha\beta} = f_{,\beta\alpha}. \tag{1.24}$$

We will return later to commutation relations of derivative operators (vectors). We will generally explicitly write out derivatives with respect to velocity or momentum components.

1.3   Tensors

We now have a class of operators, vectors, which act on functions on M. We also have a second set of operators, the 1-forms, which act on vectors. In each case the result is a real number. Moreover, a vector $\mathbf{V}$ may also be considered as an operator on 1-forms:

$$\mathbf{V}(\boldsymbol{\beta}) \equiv \boldsymbol{\beta}(\mathbf{V}) = \beta_\sigma \, \mathbf{d}\phi^\sigma (V^\alpha \boldsymbol{\partial}_\alpha) = \beta_\sigma V^\sigma. \tag{1.25}$$

Notice that $\mathbf{V}$ acts on $\boldsymbol{\beta}$ algebraically, not as a differential operator. A vector $\mathbf{V}$ is also called a *rank-1, contravariant tensor*, and a 1-form $\boldsymbol{\beta}$ is called a *rank-1, covariant tensor*. The operation shown in (1.25) is called **contraction**, and the contraction results in a number, a *rank-0 tensor*.

We now introduce the **tensor product**, which will allow us to define operators that return a real number when acting on several vectors and 1-forms. For the tensor product of two vectors $\mathbf{A}$ and $\mathbf{B}$ we write

$$\mathbf{T} = \mathbf{A} \otimes \mathbf{B}. \tag{1.26}$$

The new operation $\otimes$ is called a tensor product, and $\mathbf{T}$ is called a *rank-2, contravariant tensor*. $\mathbf{T}$ acts bilinearly and algebraically on a pair of 1-forms $\boldsymbol{\alpha}, \boldsymbol{\beta}$:

$$\mathbf{T}(\boldsymbol{\alpha}, \boldsymbol{\beta}) = \mathbf{A} \otimes \mathbf{B}(\boldsymbol{\alpha}, \boldsymbol{\beta}) \equiv \mathbf{A}(\boldsymbol{\alpha})\mathbf{B}(\boldsymbol{\beta}) = A^\sigma \alpha_\sigma B^\tau \beta_\tau. \tag{1.27}$$

In a similar way, if $\boldsymbol{\alpha}$ and $\boldsymbol{\beta}$ are 1-forms, that is, linear combinations of the basis 1-forms, one defines a *rank-2, covariant tensor* $\mathbf{R}$ by

$$\mathbf{R} = \boldsymbol{\alpha} \otimes \boldsymbol{\beta}, \tag{1.28}$$

where $\mathbf{R}$ acts bilinearly and algebraically on ordered pairs of vectors:

$$\mathbf{R}(\mathbf{A}, \mathbf{B}) \equiv \boldsymbol{\alpha}(\mathbf{A})\boldsymbol{\beta}(\mathbf{B}). \tag{1.29}$$

The **components** of a tensor are its coefficients when expanded in a basis. If the components of the two vectors and two 1-forms used above are A^μ, B^μ, α_μ, β_μ, the tensors defined above have components

$$T^{\alpha\beta} = A^\alpha B^\beta, \qquad R_{\mu\nu} = \alpha_\mu \beta_\nu. \tag{1.30}$$

It is convenient to call $T^{\alpha\beta}$ and $R_{\mu\nu}$ the tensors.

A general tensor will be a linear combination of tensor products of vectors and 1-forms. All terms must have the same number r of free upper (contravariant) indices and the same number p of free lower (covariant) indices. Such a tensor has **rank** $r + p$. It is an **r-contravariant, p-covariant tensor** or an $\binom{r}{p}$-tensor.

The definitions above were for tensors at a given point and were purely algebraic definitions. The objects can be generalized to **tensor fields**, that is, tensor-valued functions on M. Even though the underlying structure of M is differentiable, these tensor fields need not be, of course. The differentiability of a tensor field is the same as the least differentiable of its components using coordinate basis vectors and 1-forms.

Vectors, 1-forms, and tensors are geometrical objects that are defined independently of any basis system. Of course, the expression of a given vector **V** as a linear combination of basis vectors depends on which basis is used. For example, consider two sets of basis vectors $\{\bar{\mathbf{e}}_{(\alpha)}\}$ and $\{\mathbf{e}_{(\alpha)}\}$, related by

$$\bar{\mathbf{e}}_{(\alpha)} = A_{\bar\alpha}{}^\beta \mathbf{e}_{(\beta)}, \tag{1.31}$$

where $A_{\bar\alpha}{}^\beta$ is a nonsingular matrix. Note that we have used the helpful mnemonic device of writing a bar over the index $(\bar\alpha)$ that corresponds to the barred basis. The given vector **V** can be written in either basis:

$$\mathbf{V} = V^\beta \mathbf{e}_\beta = V^{\bar\alpha} \bar{\mathbf{e}}_{(\alpha)}, \tag{1.32}$$

where the components of the vector in the barred basis are written $V^{\bar\alpha}$. It immediately follows from (1.28) and (1.29) that

$$V^\beta = V^{\bar\alpha} A_{\bar\alpha}{}^\beta. \tag{1.33}$$

As an example of this transformation law for components, consider the case when both bases are partial derivative operators. The basis vectors are $\mathbf{e}_{(\beta)} = \boldsymbol{\partial}_\beta = \partial/\partial\phi^\beta$ and $\bar{\mathbf{e}}_{(\alpha)} = \boldsymbol{\partial}_{\bar\alpha} = \partial/\partial\bar\phi^\alpha$, where the second set of coordinates are $\{\bar\phi^\alpha\}$. The two sets of coordinates are presumed to be valid in the same neighborhood, and either may be considered as a function of the other, for example, $\phi^\beta = \phi^\beta(\bar\phi)$. The chain rule of differentiation implies that

$$\boldsymbol{\partial}_{\bar\alpha} = \frac{\partial\phi^\beta}{\partial\bar\phi^\alpha}\, \boldsymbol{\partial}_\beta, \quad \text{so that} \quad A_{\bar\alpha}{}^\beta = \frac{\partial\phi^\beta}{\partial\bar\phi^\alpha}. \tag{1.34}$$

Equation (1.33) may then be applied; it is then called the **coordinate transformation law** for vector components.

Similar analyses can be carried out for 1-forms or for covariant or contravariant tensors. Several easy-to-remember forms of the transformation law can be found. One of the most straightforward, for the example $S^{\alpha}{}_{\beta\gamma}$ of a $\binom{1}{2}$-tensor, is

$$S^{\alpha}{}_{\beta\gamma} A_{\bar{\nu}}{}^{\beta} A_{\bar{\lambda}}{}^{\gamma} = S^{\bar{\mu}}{}_{\bar{\nu}\bar{\lambda}} A_{\bar{\mu}}{}^{\alpha}. \tag{1.35}$$

Note that barred and nonbarred indices, both free and dummy, are matched. Other transformation laws use the matrix inverse of $A_{\bar{\mu}}{}^{\alpha}$.

The Kronecker delta is a second-rank tensor with components $\delta_{\alpha}{}^{\beta}$. Suppose that we apply a basis transformation as in (1.33) to this tensor to find the components in the barred system:

$$\delta_{\bar{\mu}}{}^{\bar{\lambda}} A_{\bar{\lambda}}{}^{\alpha} = \delta_{\beta}{}^{\alpha} A_{\bar{\mu}}{}^{\beta}$$
$$= A_{\bar{\mu}}{}^{\alpha}.$$

$A_{\bar{\lambda}}{}^{\alpha}$ is a nonsingular matrix, so, multiplying both sides on the right by $(A^{-1})_{\alpha}{}^{\bar{\sigma}}$, we obtain

$$\delta_{\bar{\mu}}{}^{\bar{\lambda}} = \delta_{\mu}{}^{\lambda}.$$

The components of the Kronecker delta do not change under frame (or coordinate) change.

An important covariant tensor is the tensor that defines **kinetic energy**. For a single, free particle moving in three-dimensional Euclidean space, this tensor is the same, up to a constant, as the metric tensor, called ds^2 (although it is not the square of anything):

$$ds^2 = \mathbf{d}x^1 \otimes \mathbf{d}x^1 + \mathbf{d}x^2 \otimes \mathbf{d}x^2 + \mathbf{d}x^3 \otimes \mathbf{d}x^3. \tag{1.36}$$

Here x^i, $i = 1, 2, 3$, are the three rectangular, Cartesian coordinates of Euclidean space. ds^2, whose components are written g_{ij}, gives the dot product of the pair of vectors on which it acts. Let $\mathbf{A} = A^i \boldsymbol{\partial}_i$ and $\mathbf{B} = B^j \boldsymbol{\partial}_j$; then

$$ds^2 = [\mathbf{d}x^1 \otimes \mathbf{d}x^1 + \mathbf{d}x^2 \otimes \mathbf{d}x^2 + \mathbf{d}x^3 \otimes \mathbf{d}x^3] \left(A^i \boldsymbol{\partial}_i, B^j \boldsymbol{\partial}_j \right)$$
$$= \sum_{\substack{i=1 \\ j=1}}^{3} \delta_{ij} A^i B^j = g_{ij} A^i B^j. \tag{1.37}$$

Thus the components of ds^2 have the numerical values $g_{ij} = \delta_{ij}$ in this rectangular coordinate frame in Euclidean space.

One important way to define a tensor is to give its components either in a single coordinate frame or in a class of such frames (in which case the various expressions must be related by the coordinate transformation law applied to these frames). In this case the components of the metric of Euclidean 3-space are defined to be δ_{ij} in any Cartesian frame. (We later discuss how to transform one Cartesian frame into another, a process that does not change the values of the components of the Euclidean metric.) The values of the components g_{ij} in other bases are determined by the component transformation law, and in general they will not be simply δ_{ij}.

Notice that the metric is an example of a symmetric tensor:

$$g_{ij} = g_{ji}. \tag{1.38}$$

This symmetry property of the components is preserved under basis transformations. A tensor $B_{\alpha\dots\gamma}$ with s indices is called completely **symmetric** if the components have the same values whenever any two indices are interchanged:

$$B_{\alpha\dots\sigma\dots\tau\dots\gamma} = B_{\alpha\dots\tau\dots\sigma\dots\gamma}.$$

Similarly, a tensor $A_{\alpha\dots\gamma}$ with r indices is called completely **antisymmetric** if the values change sign under an interchange of two indices (or in general under any odd permutation of indices):

$$A_{\alpha\dots\sigma\dots\tau\dots\gamma} = -A_{\alpha\dots\tau\dots\sigma\dots\gamma}.$$

Index symmetry properties of this sort are preserved under basis transformations, and they may be defined for completely covariant or completely contravariant tensors. To take an especially important example, the completely antisymmetric rank r tensors with all lower indices form their own subclass of geometric objects called **r -forms** and form a generalization of the 1-forms we have already introduced.

Finally, notice that symmetry properties involving one superscript and one subscript are not well defined: An equation such as $A_\mu{}^\nu = A^\nu{}_\mu$ is not a legitimate tensor equation. Thus symmetries must be defined only among indices all of which are upper or all of which are lower, for otherwise the symmetry would not be preserved under arbitrary basis transformations.

1.4 Configuration Space and Kinetic Energy

The location of a single-point particle is given by its three space coordinates x^i. A system more complex than a single-point particle needs more parameters to describe its configuration. The configuration or state of a collection of N point particles, at an instant of time t_0, is given by citing the locations of each particle. To do so, we give the coordinates of the system in a $3N$-dimensional **configuration space** R^{3N}:

$$\{x^A\} = \{x_{(1)}{}^i, x_{(2)}{}^j, \dots, x_{(N)}{}^i\}; \tag{1.39}$$

the various $x_{(a)}{}^i$ are just the locations of the ath particle, written out one after another, and the index A has the range $1, \dots, 3N$.

Time is a universal parameter in Newtonian physics and is used as the parameter of the path in configuration space that is traced out as the particles of the system move. The tangent vector is then $\mathbf{V} = d/dt$, defined as in (1.2):

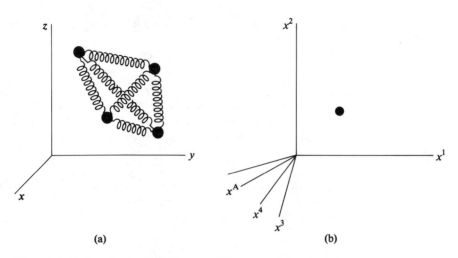

Figure 1.6. (a) Example of a physical system: Four mass points connected by springs. (b) Configuration space: 12-dimensional, with coordinates x^A. The state of the system is a point in this space $\mathbb{R}^{12}$.

We have

$$\mathbf{V} = V^A \frac{\partial}{\partial x^A} = \frac{dx^A}{dt}\, \boldsymbol{\partial}_A. \qquad (1.40)$$

At any instant t, a knowledge of the x^A and V^A suffices as initial conditions for a mechanical system (since, as we shall see, the mechanical equations are second-order, ordinary differential equations in the time), at least in ordinary cases. The $6N$-dimensional space, which consists of all possible $\{x^A, V^A\}$ taken together, that is, all possible elements of the tangent space at every possible point in the configuration space $\mathbf{R}^{3N}$, is called, as described earlier, the **tangent bundle**. [Because of Newton's 2nd Law, an element of the tangent bundle determines a history for the system.]

Since a particularly important property of a system is its **kinetic energy** KE, we must define this property of a system carefully. The kinetic energy of a single-point particle of mass m is formed by making use of the metric ds^2 of Euclidean 3-space:

$$\text{KE} = \mathbf{T}(\mathbf{V}, \mathbf{V}) = \tfrac{1}{2}m\, ds^2(\mathbf{V}, \mathbf{V}) = \tfrac{1}{2}m\delta_{ij}V^iV^j, \qquad (1.41)$$

in a Cartesian frame.

For a system of N-point particles of masses $m_{(a)}$, the total KE is the sum of the individual kinetic energies. In the language of tensors we write

$$\mathbf{T} = T_{AB}\, \mathbf{d}x^A \otimes \mathbf{d}x^B, \qquad (1.42)$$

where the components of T_{AB} may be given in diagonal matrix form, in a Cartesian frame for Euclidean 3-space, as

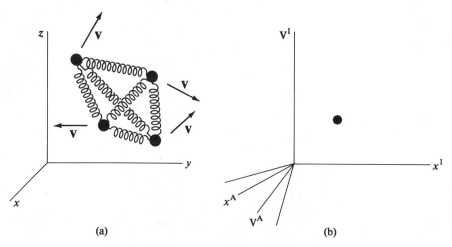

Figure 1.7. (a) The physical system with instantaneous velocities V drawn in. (b) The tangent bundle with coordinates $\{x^A, V^A\}$; it has 24 dimensions in this case: $\mathbb{R}^{24}$.

$$(T_{AB}) = \mathrm{diag}\left(\tfrac{1}{2}m_{(1)}\mathbf{I}, \tfrac{1}{2}m_{(2)}\mathbf{I}, \tfrac{1}{2}m_{(3)}\mathbf{I}, \ldots\right), \qquad (1.43)$$

where each of the blocks contains a copy of the $3\otimes3$ unit matrix $\mathbf{I}$. If a coordinate transformation is introduced in which the new coordinates are functions of the old coordinates of more than one particle, the new components T_{AB} will in general lose the diagonal form of (1.43).

1.5   Holonomic Constraints

Usually, the configuration of a system of particles is restricted by internal forces. In some cases, these restrictions, or *constraints*, take the form of functional relations among the coordinates. In such a case, the forces are called ignorable because the forces themselves never have to be considered, only the conditions on the allowed positions of the particles. Such constraints are called **holonomic constraints**.

The paradigm is a bead sliding without friction on a wire. The motion is constrained to be along the wire but is otherwise free. The forces that serve to keep the atoms of the bead together and to keep the bead on the wire are ignored (but, of course, would be investigated in other studies of the system). In this case, the position of the bead on the wire serves as the one parameter that describes the configuration of the system as a function of time. In general, many parameters are needed to describe the state of a system at an instant of time.

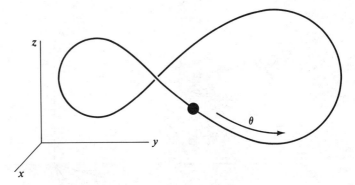

Figure 1.8. A bead constrained to move on a wire may be described either by the three coordinates of space $\{x, y, z\}$ or better by the one coordinate θ along the wire; note that configuration space is a circle here because the wire is closed.

Holonomic constraints take the form of functional relations among the coordinates:

$$f^{\mu}(x^A, t) = 0, \qquad \mu = 1, \ldots, M, \qquad (1.44)$$

where M is the number of such relations that exist. These constraints may be redundant in the sense that they are functionally related, or they may be functionally independent. We can determine precisely how many of them are functionally independent by considering the partial derivatives $\partial f^{\mu}/\partial x^A$. In particular, consider the matrix of partial derivatives $\partial f^{\mu}/\partial x^A$, the elements being in general dependent on position in $\mathbf{R}^{3N}$ and on t. A matrix is said to be of rank k if it contains at least one $k \times k$ submatrix of nonzero determinant. We assume that the rank k of $\partial f^{\mu}/\partial x^A$ is independent of position. If all the functions f^{μ} are functionally independent, then k is equal to M, the number of functions. In general, k is equal to the number of independent functions and will be less than or equal to the number of functions, and it is no more difficult to treat this more general case.

Why does this test using the rank of $\partial f^{\mu}/\partial x^A$ work? The constraints are given in terms of functions that relate one or the other coordinates. Such a (nonsingular) function f^1 restricts the solution to a one-lower-dimensional subspace, defined by $f^1 = 0$; $\partial f^1/\partial x^A$ are the components of its gradient. If a second putative constraint function has the same (or parallel, i.e., a simple multiple) gradient, it is effectively defining the same surface and adding no new constraint. The rank of the $\partial f^{\mu}/\partial x^A$ test makes use of the fact that a determinant having two rows (or two columns) proportional vanishes, so the rank is just the number of independent rows, which in our case is the number of independent constraint functions.

We thus assume that $\partial f^{\mu}/\partial x^A$ has rank k at every point in $\mathbf{R}^{3N}$ where $f^{\mu} = 0$ (for all μ) and also that the rank is k in some open set or neighborhood surrounding these points. The implicit function theorem of calculus says that the

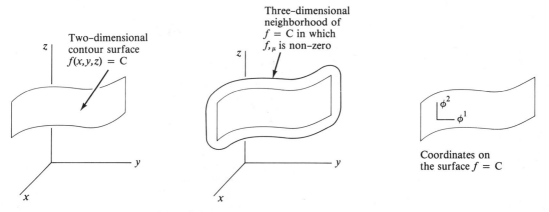

Figure 1.9. Neighborhood of $f = C$ is which $f_{,\mu}$ is nonzero. Coordinates on the surface $f = C$.

set of points such that $f^{\mu} = 0$ may be parameterized locally by $n = 3N - k$ parameters. Put another way, the implicit function theorem states that the allowed configurations, those that satisfy all constraints, form an n-dimensional space **C**, a manifold. In simplest terms, one can use the functional relations implied by the constraint functions to solve progressively for coordinates, eliminating them, eventually to obtain k of the x^A, and hence their gradients, in terms of the other (unconstrained) $n = 3N - k$. The unconstrained x^A may be taken as a possible coordinatization in one patch of **C**. Needless to say, an infinite variety of coordinate patches are possible in **C**, some of which are definitely more useful than others. It may also be the case that none of them by itself covers all of **C**. The following theorem, which we state without proof, is the justification for this procedure.

Implicit Function Theorem

    Consider the locus of points **C** *in* $\mathbf{R}^{3N}$, *which is defined by constraint functions* $f^{\mu}(x, t) = 0$, $\mu = 1, \ldots, M$. *Let* x_0^A *be a given point in* $\mathbf{C} \subset \mathbf{R}^{3N}$ *and let* $t = t_0$ *be fixed. Suppose that the rank of* $\partial f^{\mu} / \partial x^A$ *is* k *in a neighborhood of* x_0^A. *Then there exists a neighborhood* $\mathcal{N}'$ *of* x_0^A *(typically smaller than the first neighborhood) such that the points of* $\mathbf{C} \cap \mathcal{N}$ *may be uniquely parameterized by* $n = 3N - k$ *parameters* q^{α}, $\alpha = 1, \ldots, n$.

    That is, let x^A denote points in $\mathbf{C} \cap \mathcal{N}'$; then there is a 1–1 correspondence between points q^{α} in the parameter space and the points x^A. Now let t vary, and we assume that the constraints vary smoothly in time. Then we have $x^A = x^A(q, t)$.

    Because of this theorem, the functions x^A for any fixed t are functions of the q^{α} and thus are a parameterization of the points in **C** that are in a neighborhood of x_0^A. The q^{α} therefore serve as coordinates of **C** within this

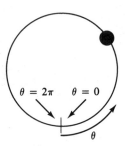

Figure 1.10. Bead on a circular wire: θ is a good coordinate except at one point.

neighborhood. We call $\mathcal{N} = \mathbf{C} \cap \mathcal{N}'$. A neighborhood $\mathcal{N}$ and coordinates q^α within $\mathcal{N}$ is a coordinate patch of $\mathbf{C}$; each point in $\mathbf{C}$ lies within such a patch. The collection of q^α, which delineates points in $\mathcal{N}$, forms an open set in $\mathbf{R}^n$, and thus the set $\mathbf{C}$ is a manifold. In fact, it is a differentiable manifold, which we presume changes smoothly as t varies.

Sometimes, however, the lack of full coverage by a particular coordinate patch may be handled by imposing boundary conditions without the necessity of using other coordinate patches. Again the paradigm is a bead sliding on a wire, in this case a circular wire. The usual angular coordinate θ can be used to cover the configuration space except for a single point: The point that would be defined by $\theta = \pi$ and $\theta = -\pi$ cannot be in the patch, for θ would not be a single-valued function there. In this case, any function to be used on this configuration space should have the periodic boundary condition

$$\lim_{\theta \to \pi} f(\theta) = \lim_{\theta \to -\pi} f(\theta), \tag{1.45}$$

and this condition obviates the necessity of using the second coordinate patch needed to cover $\mathbf{C}$.

$\mathbf{C}$ is **compact** if any open covering of $\mathbf{C}$ has a finite subcovering. Since $\mathbf{C}$ is a subset of $\mathbf{R}^{3N}$, an equivalent definition is that $\mathbf{C}$ is compact if it is a closed, bounded subset of $\mathbf{R}^{3N}$. A compact $\mathbf{C}$ is sometimes called "closed"; but that is a misleading term because closure has another meaning in topology. [Any topological space is itself both open and closed in its own topology, and any n-manifold subset of $\mathbf{R}^{3N}$ is closed (but not necessarily bounded) in the topology of $\mathbf{R}^{3N}$, if $n < 3N$.]

$\mathbf{C}$ is **connected** if any two points in $\mathbf{C}$ may be connected by a continuous path lying in $\mathbf{C}$. We will always assume that $\mathbf{C}$ is connected. $\mathbf{C}$ is said to be **simply connected** if any continuous closed path lying in $\mathbf{C}$ may be continuously shrunk to a point without leaving $\mathbf{C}$. (Other kinds of connectivity, which involve the possibility of shrinking higher dimensional subsets of $\mathbf{C}$, may also be defined.)

The restriction of the kinetic energy tensor to the ($n = 3N - k$)-dimensional hypersurface $\mathbf{C}$ becomes

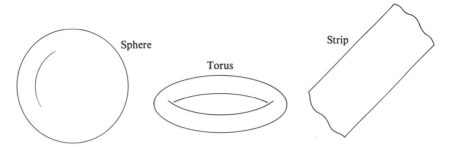

Figure 1.11. The surface of a *sphere*, the surface of a *torus* or doughnut, and the surface of a *strip* extending to infinity are all manifolds. The *strip* is noncompact, but the *sphere* and the *torus* are compact. The *sphere* is *simply connected*, as is the *strip*, but the *torus* is merely connected. A manifold which consists of any *pair* of these figures is nonconnected; we will only consider connected manifolds.

$$\mathbf{T} = T_{AB} \frac{\partial x^A}{\partial q^\alpha} \frac{\partial x^B}{\partial q^\beta} \, \mathbf{d}q^\alpha \otimes \mathbf{d}q^\beta, \tag{1.46}$$

where A, B run over $1, \ldots, 3N$, but α and β run over only the n indices associated with the coordinates q^α of $\mathbf{C}$. The x^A on which the T_{AB} depend are restricted to the values allowed by the constraints $f^\mu = 0$, which in turn implies that $x^A = x^A(q, t)$. The vector $\mathbf{V}$ on which the kinetic energy tensor acts is tangent to the hypersurface $\mathbf{C}$. This restriction automatically follows by writing

$$\mathbf{V} = V^\alpha \frac{\partial}{\partial q^\alpha} = V^\alpha \frac{\partial x^A}{\partial q^\alpha} \frac{\partial}{\partial x^A}. \tag{1.47}$$

Note that

$$\frac{dx^A}{dt} = \frac{\partial x^A}{\partial q^\alpha} \frac{dq^\alpha}{dt} + \frac{\partial x^A}{\partial t}. \tag{1.48}$$

Therefore, the kinetic energy $T_{AB}(dx^A/dt)(dx^B/dt)$ will not in general be equal to the action on $\mathbf{V}$ of the restriction of $\mathbf{T}$. If the constraints f^μ are independent of t, then $x^A = x^A(q)$ only. In this case the kinetic energy is a quadratic function of the components V^α in $\{q^\alpha\}$ coordinates. Then the kinetic energy tensor $\mathbf{T}$ has components $T_{\alpha\beta}$ given by

$$T_{\alpha\beta} = T_{AB} \frac{\partial x^A}{\partial q^\alpha} \frac{\partial x^B}{\partial q^\beta}. \tag{1.49}$$

We will mostly treat cases defined by time-independent holonomic constraints in the first chapters of this book, but the extension to time-dependent cases is not difficult. Of far greater difficulty are nonholonomic constraints or constraints where the rank of $\partial f^\mu / \partial x^A$ changes from place to place. In these cases the concept of manifold itself may not be applicable.

Examples of nonholonomic constraints include the case of a particle con-

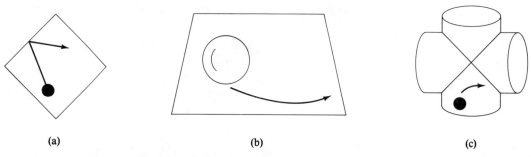

(a)   (b)   (c)

Figure 1.12. Non-Holonomic Constraints. (a) Particle in a box; (b) ball rolling on a plane; (c) particle confined to two intersecting cylinders.

fined by the hard walls of a box. Another is the case of a ball rolling without slipping on a plane. In neither case can the system's configuration be adequately described by a manifold as we are using above.

1.6 Examples

The configuration space of two particles, one allowed to move in space and the other confined to a plane, is five-dimensional Euclidean space, $\mathbf{C} = \mathbf{R}^5$. This manifold is noncompact and simply connected.

A simple pendulum is a mass m at the end of a rigid, massless rod of length R confined to swing in a plane but otherwise free to swing in a complete circle. The single point mass has Cartesian coordinates x^i ($i = 1, 2, 3$), and there are two constraint functions:

$$f^1 = (x^1)^2 + (x^2)^2 + (x^3)^2 - R^2 = 0,$$
$$f^2 = x^2 = 0.$$

The constraint $f^2 = 0$ confines the particle to the $(x^1$–$x^3)$-plane. The constraint $f^1 = 0$ constrains the particle to the sphere with radius R centered on the origin. The combined constraints mean that configuration space is a circle (or one-dimensional sphere) S^1. This manifold is compact but not simply connected.

The spherical pendulum is a point mass constrained by the single constraint $f^1 = 0$. The configuration space $\mathbf{C}$ is the ordinary sphere S^2 (called S^2, since it is a two-dimensional surface), which is compact and simply connected.

The simple pendulum may also be described as a point mass subject to the constraint $f^2 = 0$ and another constraint,

$$f^3 = (x^1)^2 + (x^3)^2 - R^2 = 0.$$

All three constraints $f^1 = f^2 = f^3 = 0$ may be used. In this case the 3×3 matrix of derivatives $\partial f^\mu / \partial x^i$ is

$$\frac{\partial f^\mu}{\partial x^i} = \begin{pmatrix} 2x^1 & 2x^2 & 2x^3 \\ 0 & 1 & 0 \\ 2x^1 & 0 & 2x^3 \end{pmatrix}.$$

This matrix has rank 2 unless $x^1 = x^3 = 0$, but the latter situation cannot occur when $f^1 = f^2 = f^3 = 0$. Therefore, $n = 3N - k = 1$, as before.

A particle confined to the $(x^1$–$x^3)$-plane may be defined as a single particle subject to the constraint $f^2 = x^2 = 0$. The 1×3 matrix $\partial f^\mu / \partial x^i$ has rank 1. The constraint $(x^2)^2 = 0$, however, although apparently the same, should not be used, since $\partial f^\mu / \partial x^i$ in this case has all zero entries when $x^2 = 0$. On the other hand, any other constraint function $f(x^2) = 0$, which has the single solution $x^2 = 0$ and which has nonvanishing derivative when $x^2 = 0$, would be fine.

Our examples of manifolds have been given in a coordinate-free manner in the sense that we did not specify coordinates in the manifolds themselves. Finding convenient coordinates is not always simple. Consider S^2, defined by

$$(x^1)^2 + (x^2)^2 + (x^3)^2 - R^2 = 0.$$

Call the points where $x^3 = \pm R$ the north and south poles. Around the north pole x^1 and x^2 may be used as coordinates, since x^3 may be defined uniquely as $+ [R^2 - (x^1)^2 - (x^2)^2]^{1/2}$. However, as coordinates on S^2, x^1 and x^2 cannot be used beyond the $x^3 = 0$ equator.

The usual spherical coordinates r, θ, ϕ are defined by

$$x^1 = r \sin \theta \cos \phi,$$
$$x^2 = r \sin \theta \sin \phi,$$
$$x^3 = r \cos \theta,$$

and the sphere of radius R is defined by $r = R$. As coordinates on S^2, θ and ϕ have the ranges $0 < \theta < \pi$ and $-\pi < \phi < \pi$. They are useful on all of S^2 except for the line defined by $\phi = \pi$, namely, the line from the north pole to the south pole along the international date line. As we mentioned above, for most purposes a single coordinate system is adequate. As in the case of the circular wire, the coordinates θ, ϕ on S^2 can be used provided appropriate boundary conditions are employed; then it is not in fact necessary to provide a second coordinate patch around the line $\phi = \pi$.

A pathological situation is that of a particle simultaneously required to lie on a cylinder surrounding the x^1-axis and on a cylinder surrounding the x^2-axis, both cylinders having the same radius. In the latter case, configuration space is a pair of intersecting closed curves; the two points of intersection show that **C** is not a manifold in this case.

EXERCISES

1.1. Let $S^{\alpha\beta} = S^{\beta\alpha}$. Let $A_{\alpha\beta} = -A_{\beta\alpha}$. Let $S^{\alpha\cdots\gamma}$ be completely symmetric and of rank s. Let $A_{\alpha\cdots\gamma}$ be completely antisymmetric and of rank r. The dimension of the manifold on which these tensors are defined is n.
(a) Show that $S^{\alpha\beta} A_{\alpha\beta} = 0$.
(b) Find the maximum number of linearly independent components of $S^{\alpha\beta}$ and $A_{\alpha\beta}$.

(c) Suppose that $r > n$. Show that $A_{\alpha\ldots\gamma} = 0$.

(d) Suppose that $r = n$. Show that $A_{\alpha\ldots\gamma}$ has exactly one independent component.

(e) Find the number of independent components of $A_{\alpha\ldots\gamma}$ for general r.

(f) Find the number of independent components of $S_{\alpha\ldots\gamma}$ for general s.

(g) A reduced symmetric tensor $\overset{\circ}{S}{}^{\alpha\ldots\gamma}$ is one which, in addition to being completely symmetric, has zero trace, the trace being defined by use of a tensor $g_{\alpha\beta}$ (a metric tensor, one whose components form a nonsingular matrix in the sense that $\det|g_{\alpha\beta}| \neq 0$):

$$\overset{\circ}{S}{}^{\alpha\ldots\beta\ldots\gamma}g_{\alpha\beta} = 0.$$

Find the number of independent components of $\overset{\circ}{S}{}^{\alpha\ldots\gamma}$ for general rank s.

1.2. As mentioned in the text, one way of defining a tensor is to give its components in one coordinate system, assuming that the manifold $\mathcal{M}$ can be covered by a single coordinate system $\{\phi^\alpha\}$. An important example is the completely antisymmetric rank-n covariant tensor ε whose components satisfy $\varepsilon_{1\ldots n} = +1$. Find the components of ε in new coordinates $\{\bar{\phi}^\alpha\}$. (*Hint:* Show that ε has only one independent component and that one can be found if you know $\det|\partial\phi^\beta/\partial\bar{\phi}^\alpha|$.)

1.3. Let $C_{\alpha\ldots\gamma}$ be a rank r tensor, not with any particular symmetry. Suppose that $C_{\alpha\ldots\gamma}D^{\alpha\ldots\gamma} = 0$ for any tensor $D^{\alpha\ldots\gamma}$. Prove that $C_{\alpha\ldots\gamma} = 0$.

1.4. Let $\mathbf{V}$ and $\mathbf{W}$ be two vectors $\mathbf{V} = V^\alpha\partial_\alpha$, $\mathbf{W} = W^\alpha\partial_\alpha$. In acting on a function f, $\mathbf{VW}$ is a second-order derivative operator. Let $[\mathbf{V}, \mathbf{W}] \equiv \mathbf{VW} - \mathbf{WV}$ be the **commutator** of $\mathbf{V}$ and $\mathbf{W}$ as differentiation operators. Show that $[\mathbf{V}, \mathbf{W}]$ is a vector (i.e., it is a first-order differentiation operator). Also, show that if $\mathbf{U}$ is a third vector, the **Jacobi identity** holds:

$$[[\mathbf{U}, \mathbf{V}], \mathbf{W}] + [[\mathbf{V}, \mathbf{W}], \mathbf{U}] + [[\mathbf{W}, \mathbf{U}], \mathbf{V}] = 0.$$

1.5. In an n-dimensional space $\mathbf{C}$, the solution of an equation $f(\phi) = 0$ in general defines an $(n-1)$-dimensional subspace $\mathbf{D}$ in $\mathbf{C}$. Show that the 1-form $f_{,\mu}$ is orthogonal to $\mathbf{D}$. That is, let $\phi^\alpha(t)$ be a curve lying in $\mathbf{C}$, which also lies in $\mathbf{D}$. Let $d\phi^\alpha/dt$ be the velocity of the curve; by definition, $d\phi^\alpha/dt$ is tangent to the curve and is a generic tangent vector to $\mathbf{D}$. Show that

$$f_{,\alpha}\frac{d\phi^\alpha}{dt} = 0.$$

1.6. The 3-sphere S^3 is the locus of points in $\mathbf{R}^4$ defined by ($R = $ const)

$$F = (x^1)^2 + (x^2)^2 + (x^3)^2 + (x^4)^2 - R^2 = 0.$$

Define three vectors $\mathbf{A}$, $\mathbf{B}$, and $\mathbf{C}$ by

$$\mathbf{A} = x^2\partial_1 - x^1\partial_2 + x^4\partial_3 - x^3\partial_4,$$

$$\mathbf{B} = -x^3 \partial_1 + x^4 \partial_2 + x^1 \partial_3 - x^2 \partial_4,$$
$$\mathbf{C} = -x^4 \partial_1 - x^3 \partial_2 + x^2 \partial_3 + x^1 \partial_4.$$

Show that $\mathbf{A}$, $\mathbf{B}$, and $\mathbf{C}$ are all tangent to S^3. Compute (see Exercise 1.4) $[\mathbf{A}, \mathbf{B}]$, $[\mathbf{B}, \mathbf{C}]$, and $[\mathbf{C}, \mathbf{A}]$. Find three 1-forms μ, ν, and λ whose components are $\mu = \mu_\alpha \, \mathbf{d}x^\alpha$, $\nu = \nu_\alpha \, \mathbf{d}x^\alpha$, $\lambda = \lambda_\alpha \, \mathbf{d}x^\alpha$, such that

$$\mu(\mathbf{A}) = 1, \qquad \mu(\mathbf{B}) = 0, \qquad \mu(\mathbf{C}) = 0,$$
$$\nu(\mathbf{A}) = 0, \qquad \nu(\mathbf{B}) = 1, \qquad \nu(\mathbf{C}) = 0,$$
$$\lambda(\mathbf{A}) = 0, \qquad \lambda(\mathbf{B}) = 0, \qquad \lambda(\mathbf{C}) = 1.$$

Note that μ, ν, and λ are not unique because any multiple of $\mathbf{d}F$ may be added to any one of them without affecting the relations above.

1.7. The surface of a 2-sphere S^2 is a manifold, and in the text coordinate systems were given, but only for part of it. Complete the proof that this surface is a manifold by showing that any point can be covered by a coordinate patch.

1.8. Is the surface of a cone a differentiable manifold? (*Hint:* Consider the neighborhoods of the point at the vertex of the cone.)

1.9. A possible coordinate system on the 2-plane $\mathbf{R}^2$ is the cylindrical system, with r the distance from the origin and ϕ the angle measured counterclockwise from the x-axis (x and y being the usual Cartesian coordinates).
(a) Show that the transformation from the rectangular coordinates $\{x, y\}$ to $\{r, \phi\}$ is nonsingular everywhere except on a half-line that includes the point $x = y = 0$. (Thus the $\{r, \phi\}$ coordinate patch is smaller than the $\{x, y\}$ one, which covers the whole plane.)
(b) What are the coordinate basis vectors ∂_α in the $\{r, \phi\}$ patch, expressed in terms of the basis $\{\partial/\partial x, \partial/\partial y\}$?
(c) What are the 1-forms dual to the r, ϕ basis $\{\partial_\alpha\}$, expressed in terms of the basis $\{\mathbf{d}x, \mathbf{d}y\}$?
(d) Using the results from part (c), write the two-dimensional metric tensor ds^2 in terms of $\{\mathbf{d}r, \mathbf{d}\phi\}$.
(e) It is often useful to work with unit vectors in developing physical results. [A vector $\mathbf{V}$ is unit if $ds^2(\mathbf{V}, \mathbf{V}) = 1$.] Using the results of part (d), show that the basis vectors $\{\partial_\alpha\}$ in the r and ϕ directions are orthogonal. Show also that a simple rescaling makes them unit as well, so that one can define a basis of vectors $\{\mathbf{e}_{(\alpha)}\}$ that is orthonormal.
(f) Using Exercise 1.4, show that the commutator of the orthonormal vectors found in part (e) is $[\mathbf{e}_\phi, \mathbf{e}_r] = r^{-1}\mathbf{e}_\phi$.

1.10. (a) Calculate the differential $\mathbf{d}h$ of the function

$$h(x, y, z) = 3x^3 y + yz + xz^2,$$

where x, y, z are rectangular coordinates.

(b) Consider the vector

$$\mathbf{V} = x^5 \boldsymbol{\partial}_1 + 2yzx^2 \boldsymbol{\partial}_2 + xy^2 \boldsymbol{\partial}_3,$$

where $\boldsymbol{\partial}_i$ are the rectangular coordinate basis vectors. Evaluate the expression $\mathbf{d}h(\mathbf{V})$ at the point $(x, y, z) = (2, 1, 3)$.

1.11. (a) A differential form is given in spherical coordinates by

$$\boldsymbol{\omega} = \sin^2 \theta \, \mathbf{d}r + 2r \sin \theta \cos \theta \, \mathbf{d}\theta + r \sin^2 \theta \, \mathbf{d}\phi.$$

Is this the differential of a particular function? If so, give the function Q such that $\boldsymbol{\omega} = \mathbf{d}Q$. If not, explain why not. (*Hint:* If $\boldsymbol{\omega}$ is the gradient of a function Q, then $\omega_\alpha = \boldsymbol{\partial}_\alpha Q$.)

(b) This is the same problem as in part (a), except use the 1-form

$$\boldsymbol{\omega} = x \, \mathbf{d}y + z \, \mathbf{d}x,$$

$\{x, y, z\}$ being the usual rectangular coordinate system.

1.12. Let $\mathcal{N} \subset S^2$ be the set that includes all points except the north pole A. Place the sphere S^2 so that its south pole sits at the origin of $\mathbf{R}^2$ (Cartesian coordinates $\{x, y\}$). Draw a straight line from A to a point (x, y) of $\mathbf{R}^2$; let the point where it intersects S^2 be called P. The numbers x, y thus are coordinates in $\mathcal{N}$. Suppose that P is the position of a particle confined to move on S^2 but otherwise unconstrained. Find the kinetic energy tensor of the particle in terms of these coordinates.

1.13. A diatomic solid consists of two atoms a fixed distance apart but otherwise free to move. Describe the configuration space $\mathbf{C}$, give its dimension, and tell whether it is compact or simply connected. Give coordinates that cover at least part of $\mathbf{C}$.

1.14. A simple system consists of eight atoms arranged in a cube, and the distance between any two is fixed. There are thus $\frac{1}{2}(8 \times 7) = 28$ constraint functions. What are they? What is the dimension n of configuration space $\mathbf{C}$? If you are a sucker for punishment, try writing out all 28 constraints f^μ, and try finding by direct calculation the rank of $\partial f^\mu / \partial x^A$.

1.15. A square consists of four atoms confined to lie in a rigid square array in the $(x^1 – x^2)$-plane; the distance between any two atoms is fixed. It is easy to reduce the $3N = 12$ spatial coordinates to eight, the x^1, x^2 coordinates of the four atoms in the plane. Show that there are six additional constraints f^μ. Show that the 6×8 matrix $\partial f^\mu / \partial x^A$ (here $A = 1, \ldots, 8$) has rank 5 (at least near the system point with the square centered on the origin and oriented with one atom on the x^1-axis) by direct calculation: Consequently, configuration space $\mathbf{C}$ is three-dimensional. Is $\mathbf{C}$ compact or simply connected?

1.16. What is the configuration space for a solid body with one point fixed in space? The body consists of $N \geq 4$ noncoplanar particles such that the distance between any two particles is fixed. Aside from the three constraints that fix the location of one particle, there are $N(N-1)/2$ other constraint functions. The total number of constraints is therefore $3 + [N(N-1)/2]$, which is $\geq 3N-3$. However, show that $n = 3N - k$ [where k is the rank of $(\partial f^\mu / \partial x^A)$] is 3. Is **C** compact or simply connected?

1.17. Consider a particle required to lie simultaneously on two equal cylinders, one surrounding the x^1-axis and the other surrounding the x^2-axis. The two constraint functions are (R is constant)

$$f^1 = (x^2)^2 + (x^3)^2 - R^2 = 0,$$
$$f^2 = (x^1)^2 + (x^3)^2 - R^2 = 0.$$

Compute the rank k of $(\partial f^\mu / \partial x^i)$, and show that there are two points in configuration space **C** where the rank is not 2. Draw a picture of **C** and show that the existence of these two points keep **C** from being a manifold.

1.18. Let a particle be confined to the surface of a sphere S^2 of radius R. In the northern hemisphere its three spatial coordinates are given in terms of two variables q^1, q^2 by

$$x^1 = q^1,$$
$$x^2 = q^2,$$
$$x^3 = +[R^2 - (q^1)^2 - (q^2)^2]^{1/2}.$$

Find $T_{\alpha\beta}$, where $\alpha, \beta = 1, 2$.

1.19. A particle free to move in space (without constraint) may be described in Cartesian coordinates $\{x^i\}$. If cylindrical coordinates $\rho = q^1, \theta = q^2, \zeta = q^3$ are used, then

$$x^1 = \rho\cos\theta, \qquad x^2 = \rho\sin\theta, \qquad x^3 = \zeta.$$

If spherical coordinates $r = q^1, \theta = q^2, \phi = q^3$ are used, then

$$x^1 = r\sin\theta\cos\phi, \qquad x^2 = r\sin\theta\sin\phi, \qquad x^3 = r\cos\theta.$$

In both cases compute $T_{\alpha\beta}$, where $\alpha, \beta = 1, 2, 3$.

1.20. Let L be a curve $q^\alpha(t)$ in configuration space **C**. If the physical system has no potential energy and if the kinetic energy is of the quadratic type described in this chapter, the momentum p_α for this curve is defined by

$$p_\alpha = 2T_{\alpha\beta}\frac{dq^\alpha}{dt}.$$

A second curve L' has a different path parameter t' but traces out the same orbit in the sense that for every value of t' there is a value of t such that $q^\alpha(t) = q^\alpha(t')$. Suppose it happens that the curve L satisfies

$$\frac{dp_\alpha}{dt} = T_{\sigma\tau,\alpha}\frac{dq^\sigma}{dt}\frac{dq^\tau}{dt},$$

where the comma means partial derivative, as in (1.18). Suppose it also happens that the curve L' satisfies this equation in the form

$$\frac{d}{dt'}\left(T_{\alpha\beta}\frac{dq^{\beta}}{dt'}\right) = T_{\sigma\tau,\alpha}\frac{dq^{\sigma}}{dt'}\frac{dq^{\tau}}{dt'}.$$

Show that t and t' are related by a linear transformation. Also, show that the kinetic energy $T_{\sigma\tau}(dq^{\sigma}/dt)(dq^{\tau}/dt)$ is constant along L (presuming that $\partial T_{\alpha\beta}/\partial t = 0$).

1.21. (This is a generalization of Exercise 1.5.) In the space $\mathbf{R}^{3N}$ with coordinates x^A, let $\mathbf{C}$ be defined by the equations $f^{\mu} = 0$, $\mu = 1, \ldots, n$. A path lying in $\mathbf{C}$ is defined by the one-dimensional subset $x^A(t)$, provided that $f^{\mu}(x^A(t)) = 0$. Let P_0 be a point in $\mathbf{C}$, and let L be any path passing through P_0 and lying in $\mathbf{C}$. Consider the $3N$-tuple of numbers

$$z_A \equiv K_{\mu}f^{\mu}{}_{,A}(P_0) \qquad \text{(evaluated at } P_0\text{)},$$

where $K_{\mu} = \text{const}$. Show that if V^A is the tangent to L at P_0, then $z_A V^A = 0$. Conclude that z_A is orthogonal to $\mathbf{C}$ in the sense that z_A is orthogonal to the tangent vector of any path in $\mathbf{C}$ passing through P_0. Prove that the set of all linear combinations of possible z_A's is a k-dimensional vector space if the matrix $(f^{\mu}{}_{,A})$ has rank k. The vectors of this space are all orthogonal to $\mathbf{C}$, and hence $\mathbf{C}$ has dimension $3N - k$.

1.22. Consider a system of two particles, P_1 and P_2. P_1 is confined to a unit-radius circular track in the $z = 0$ plane, centered at $x = y = 0$. Particle P_2 is confined to a unit-radius circular track in the $z = 1$ plane centered at $x = y = 0$. (P_1 and P_2 are to be connected by a spring with force constant k, but this fact will only be used in later exercises, in later chapters.) Describe the two-dimensional configuration space $\mathbf{C}$ of this system. Prove that $\mathbf{C}$ is a manifold in the following manner: About each point in $\mathbf{C}$ find a continuous 1–1 map ϕ, which maps some neighborhood of the point onto an open subset of $\mathbf{R}^2$.

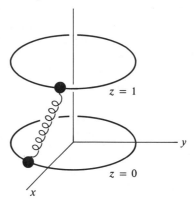

Figure P1.22. Two particles, confined to two circular tracks, connected by a spring.

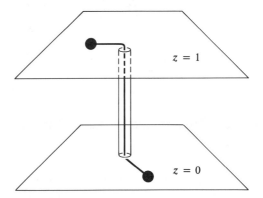

Figure P1.23. Particles confined to two planes. They are connected by an inextendible string through a thin tube.

1.23. A similar but more interesting problem is to describe the configuration space of two particles related in the following way: Particle P_1 is confined to the $z = 1$ plane, and particle P_2 is confined to the $z = 0$ plane. These two planes are connected to each other by a thin, hollow tube along the z-axis. Through this tube is a light, inextendible string, which connects the particles. Therefore, the sum of the distances of the two particles from their respective origins is a constant.

1.24. As in Exercise 1.6, let the three-sphere S^3 be defined as the locus of points in $\mathbf{R}^4$ subject to the constraint ($R = \text{const.}$)

$$F = \sum_{A=1}^{4} (x^A)^2 - R^2 = 0.$$

Obviously, the constraint can be solved for x^4 as a single-valued function of x^1, x^2, x^3 in the upper-half sphere, where $x^4 > 0$. Introduce parameters ψ, θ, ϕ to serve as coordinates on S^3, where

$$x^1 = R \sin\psi \sin\theta \cos\phi,$$
$$x^2 = R \sin\psi \sin\theta \sin\phi,$$
$$x^3 = R \sin\psi \cos\theta,$$

and use the constraint F to determine what functional dependence x^4 has on ψ, θ, ϕ. Then assume that the metric in $\mathbf{R}^4$ is $ds^2 = \delta_{AB}\, \mathbf{d}x^A \otimes \mathbf{d}x^B$. Calculate the metric on S^3 in terms of the basis $\{\mathbf{d}\psi, \mathbf{d}\theta, \mathbf{d}\phi\}$.

2

SPACETIME

The study of time-dependent constraints in Newtonian physics is best carried out by using the concept of spacetime from relativity. Some modification—actually a simplification—in the properties of spacetime is necessary since time in Newtonian mechanics is a special parameter. Spacetime is the manifold $\mathcal{M}$ of all events. (We use $\mathcal{M}$ for the four-dimensional spacetime of a single particle, since that is a commonly used symbol for Minkowski spacetime in special relativity. However, we do not use the relativity *metric* in this chapter.) Each point must be specified by giving a spatial location and a time; we call a point of $\mathcal{M}$ an event even if nothing happens to have occurred there.

$\mathcal{M}$ is four-dimensional. The history of a single particle, its worldline, is a parameterized path through $\mathcal{M}$. Let the spatial position of the particle be $x^i(t)$; its path through $\mathcal{M}$ is then the quadruple $(x^i(t), t)$. The histories of N particles are N different paths in $\mathcal{M}$:

$$(x^i{}_{(a)}, t_{(a)}) \qquad \text{for} \quad i = 1, 2, 3 \quad \text{and} \quad a = 1, \ldots, N. \qquad (2.1)$$

To preserve the special Newtonian nature of time, however, we will assume that

$$t_{(1)} = t_{(2)} = \cdots = t_{(N)}. \qquad (2.2)$$

Spacetime for N particles is thus $(3N + 1)$-dimensional.

We denote this spacetime by $\mathbf{R}^{3N} \times \mathbf{R}$ to emphasize that the last $\mathbf{R}$ represents the real line of all possible times. In general, if we have a manifold $\mathbf{C}$, with points denoted q^i in some coordinate system, the manifold $\mathbf{C} \times \mathbf{R}$ is the collection of all pairs consisting of a point in $\mathbf{C}$ and a number from $\mathbf{R}$. The points in $\mathbf{C} \times \mathbf{R}$ may thus be denoted (q^i, t), where q^i is in $\mathbf{C}$ and t is in $\mathbf{R}$.

Return now to the spacetime of N free particles $\mathbf{R}^{3N} \times \mathbf{R}$. Constraints within spacetime are of the form

$$f^\mu(x^A, t) = 0, \qquad \mu = 1, 2, \ldots, \tag{2.3}$$

where, as before, the spatial dependent variables are x^A, $A = 1, \ldots, 3N$. If the matrix of derivatives

$$\left(\frac{\partial f^\mu}{\partial x^A}, \frac{\partial f^\mu}{\partial t} \right) \tag{2.4}$$

has rank k everywhere, spacetime reduces to a manifold $\mathbf{D}$ of $3N + 1 - k = n + 1$ dimensions. Consider the intersection of $\mathbf{D}$ with a $t = $ constant spatial slice of $\mathbf{R}^{3N} \times \mathbf{R}$; we call the intersection $\mathbf{C}_t$:

$$\mathbf{C}_t = \mathbf{D} \cap \mathbf{R}^{3N}. \tag{2.5}$$

$\mathbf{C}_t$ need not have any especially tractable properties, and it is easy to give examples where $\mathbf{C}_t$ changes topology at various times. However, as is usual, we will adopt the assumption that $\partial f^\mu / \partial x^A$ by itself has rank k and that all of the resulting n-dimensional $\mathbf{C}_t$'s can be put into a 1–1 (and differentiable) correspondence with a single manifold $\mathbf{C}$.

Configuration spacetime, with the assumption above, is the $(n + 1)$-dimensional manifold $\mathbf{C} \times \mathbf{R}$. A point in this manifold is (q^i, t) in a given coordinate system, so that the solution of $f^\mu(x^A, t) = 0$ is given by

$$\begin{aligned} x^A &= x^A(q^i, t) \quad \text{with} \quad A = 1, \ldots, 3N, \quad i = 1, \ldots, n, \\ t &= t. \end{aligned} \tag{2.6}$$

We have added the last equation to indicate that $\mathbf{C} \times \mathbf{R}$ is a subset of $\mathbf{R}^{3N} \times \mathbf{R}$ requiring that the time t be a special coordinate. Of course, the $\{q^i\}$ coordinates may, as before, only be applicable in a limited region in $\mathbf{C}$.

A change of coordinates in spacetime in principle allows time t to be changed to a new parameter $\bar{t}$. We thus have

$$\begin{aligned} \bar{q}^j &= \bar{q}^j(q, t) \quad \leftrightarrow \quad q^i = q^i(\bar{q}, \bar{t}), \\ \bar{t} &= \bar{t}(q, t) \quad \leftrightarrow \quad t = t(\bar{q}, \bar{t}). \end{aligned} \tag{2.7}$$

To preserve the special Newtonian nature of time, however, we will assume that

$$\bar{t} = t. \tag{2.8}$$

Now let us consider a path in spacetime $\big(q^i(t),t\big)$ in a particular coordinate system. The spacetime velocity of this path is the differential operator d/dt, where t is considered as the path parameter. More precisely, let us call the path parameter τ and define the spacetime path by

$$\big(q^i(\tau),t(\tau)\big), \qquad \text{where} \quad t(\tau) = \tau. \tag{2.9}$$

The spacetime velocity is thus properly written as $d/d\tau$, and it has components

$$\left(\frac{dq^i}{d\tau}, \frac{dt}{d\tau}\right). \tag{2.10}$$

Since (again) Newtonian physics always uses t as a parameter: $t = \tau$, we can write these components as

$$(\dot{q}^i, 1).$$

A general vector in spacetime has components (v^i, v^0), where v^0 is the component in the t-direction. In the language of partial derivatives this vector is

$$\frac{d}{d\tau} \equiv v^i \frac{\partial}{\partial q^i} + v^0 \frac{\partial}{\partial t}. \tag{2.11}$$

Be very careful not to confuse $\partial/\partial t$ with d/dt. The notation d/dt means derivative with respect to t as a path parameter and is more correctly defined, as above, as $d/d\tau$ with $t = \tau$.

Under a change of coordinates to $\{\bar{q}, \bar{t}\}$, the components of the spacetime vector change to $(v^{\bar{j}}, v^{\bar{0}})$ where

$$\begin{aligned}
v^{\bar{j}} &= v^i \frac{\partial \bar{q}^j}{\partial q^i} + v^0 \frac{\partial \bar{q}^j}{\partial t}, \\
v^{\bar{0}} &= v^i \frac{\partial \bar{t}}{\partial q^i} + v^0 \frac{\partial \bar{t}}{\partial t}.
\end{aligned} \tag{2.12}$$

We do assume that $\bar{t} = t$. In this case $(v^{\bar{j}}, v^{\bar{0}})$ is given by

$$\begin{aligned}
v^{\bar{j}} &= v^i \frac{\partial \bar{q}^j}{\partial q^i} + v^0 \frac{\partial \bar{q}^j}{\partial t}, \\
v^{\bar{0}} &= v^0.
\end{aligned} \tag{2.13}$$

Notice that v^0 does not change if $\bar{t} = t$. The velocity vector $(\dot{q}^i, 1)$ of a path therefore retains the constant **one** in the time component under a change in spacetime coordinates, which preserves t. Moreover, this result is the reason why configuration space $\mathbf{C}$ by itself can be used, rather than spacetime, even when time-dependent constraints are used in its definition. To be specific, let v^A be the components of the velocity considered as a vector in $\mathbf{R}^{3N}$:

$$v^A = \frac{d}{dt}.$$

Since $x^A = x^A(q,t)$ is the solution of the time-dependent constraints, we may write

$$v^A = \frac{\partial x^A}{\partial q^i}\frac{dq^i}{dt} + \frac{\partial x^A}{\partial t}. \tag{2.14}$$

This formula is the one that would be written in a fairly naive application of the chain rule for derivatives. It is, of course, the same as the previously written formula (2.12) for the change of coordinates with $\bar{q}^i$ replaced by x^A and with $v^{\bar{j}}$ replaced by v^A. The spacetime meaning of this formula is the proper one, however.

In practice the use of

$$v^A = \frac{\partial x^A}{\partial q^i}\dot{q}^i + \frac{\partial x^A}{\partial t}$$

is straightforward. Consider, for example, kinetic energy, defined in $\mathbf{R}^{3N}$ by

$$T = T_{AB}\,\dot{x}^A\dot{x}^B.$$

In $\mathbf{C}$, the expression for kinetic energy is given by substituting for $\dot{x}^A = v^A$:

$$T = T_{AB}\left(\frac{\partial x^A}{\partial q^i}\dot{q}^i + \frac{\partial x^A}{\partial t}\right)\left(\frac{\partial x^B}{\partial q^j}\dot{q}^j + \frac{\partial x^B}{\partial t}\right). \tag{2.15}$$

The result has a term quadratic in $\dot{q}^i$, a term linear in $\dot{q}^i$, and a term independent of $\dot{q}^i$:

$$T = T_{ij}\,\dot{q}^i\dot{q}^j + S_i\,\dot{q}^i + R, \tag{2.16}$$

where

$$T_{ij} = T_{AB}\frac{\partial x^A}{\partial q^i}\frac{\partial x^B}{\partial q^j},$$

$$S_i = 2\,T_{AB}\frac{\partial x^A}{\partial q^i}\frac{\partial x^B}{\partial t}, \tag{2.17}$$

$$R = T_{AB}\frac{\partial x^A}{\partial t}\frac{\partial x^B}{\partial t}.$$

In the exercises we will indicate how T is computed as a quantity pertaining to configuration spacetime $\mathbf{C}$.

EXERCISES

2.1. A free particle is confined to the $x^3 = 0$ plane. Its equations of motion are therefore $\ddot{x}^A = 0$, $A = 1, 2$. These equations may also be expressed as

$$\frac{d}{dt}\frac{\partial T}{\partial \dot{x}^A} = \frac{\partial T}{\partial x^A}.$$

Uniformly rotating coordinates q^i, $i = 1, 2$, are defined by

$$x^1 = q^1 \cos \omega t - q^2 \sin \omega t,$$
$$x^2 = q^1 \sin \omega t + q^2 \cos \omega t.$$

Express T in the q^i coordinate system. Show what the equations $\ddot{x}^A = 0$ become in terms of the q^i coordinates. Show that these equations are expressible as

$$\frac{d}{dt}\frac{\partial T}{\partial \dot{q}^i} = \frac{\partial T}{\partial q^i}.$$

2.2. A particle is confined to a frictionless plane containing the x^3-axis. This plane rotates with uniform angular speed ω about the x^3-axis. Find T in suitable coordinates.

2.3. Consider a tensor in $\mathbf{R}^{3N} \times \mathbf{R}$ (spacetime) that has components including the T_{AB} of ordinary kinetic energy. It will have $(3N + 1)^2$ components, and these can be arranged in a symbolic square array. The column and row involving a time component will result in the components labeled T_{0A} or T_{A0}, but we will take these components to be zero. The T_{00} component is the component when both tensor indices correspond to the time direction. We take the component to be a constant $-E$. The components of the $(3N + 1) \times (3N + 1)$ tensor are then

$$\left[\begin{array}{c|c} T_{AB} & 0 \\ \hline 0 & -E \end{array}\right].$$

In $\mathbf{C} \times \mathbf{R}$ configuration spacetime, the components form an $(n+1) \times (n+1)$ array (where $\bar{T}_{0i} = \bar{T}_{i0}$):

$$\left[\begin{array}{c|c} \bar{T}_{ij} & \bar{T}_{i0} \\ \hline \bar{T}_{0j} & \bar{T}_{00} \end{array}\right].$$

Find this array. The quantity corresponding to kinetic energy is $\bar{T}$ given by

$$\bar{T} = \bar{T}_{ij} v^i v^j + 2\,\bar{T}_{0i} v^0 v^i + \bar{T}_{00} v^0 v^0,$$

where

$$v^j = \dot{q}^j, \qquad v^0 = 1.$$

Find $\bar{T}$. [*Note:* Do not assume time independence of the constraints. Thus $x^A = x^A(q, t)$ may explicitly depend on t.]

3

LAGRANGIANS AND LAGRANGIAN SYSTEMS

Thus far we have seen how to write the kinetic energy in a geometrical way, taking advantage of the fact that the kinetic energy is a tensor on the tangent bundle. In conservative systems, there is in addition a scalar function, the potential energy $V(x^A)$. Newton's equations in their second-order form are vector equations, which relate accelerations to gradients of this potential energy. Because of their vector nature, these equations are difficult in practice (but not in principle!) to transform from one basis set to another.

In fact, these equations can be encoded into a single scalar function L on the tangent bundle (i.e., a function of the coordinates, of the velocities, and of time):

$$L = L(x^A, \dot{x}^A, t). \tag{3.1}$$

The function L is called the Lagrangian. Its definition for simple conservative systems (e.g., for the description of point mass dynamics in conservative fields) is

$$L = T - V, \tag{3.2}$$

where T is the kinetic energy and V is the potential energy.

Now consider the **action** associated with the Lagrangian (3.1):

$$I = \int_{t_a;C}^{t_b} L(x^A, \dot{x}^A, t)\, dt, \tag{3.3}$$

where t_a and t_b are two fixed times, and $C: \mathbf{R} \to \mathcal{M}$ is a curve. The curve C is a parameterized path where $t \to x^A(t)$, and it gives the time behavior of $x^A(t)$ in the integrand and hence the velocity components $\dot{x}^A = dx^A/dt$. The endpoints of the path, whose coordinates are $x^A(t_a)$ and $x^A(t_b)$, are considered as fixed.

The algorithm that allows us to extract Newton's equations of motion from the Lagrangian is the following: *I is an extremum compared to curves infinitesimally differing from C (with the same fixed endpoints) if and only if C specifies $x^A(t)$, which is a solution to Newton's equations.*

Let us investigate the properties of I near its extreme value. We suppose that $C: \mathbf{R} \to \mathcal{M}|t \to x^A(t)$ is the curve that gives the extremum. Adjacent curves with the same endpoints can be parameterized by adding a small, smooth vector function of t to the function $x^A(t)$ that specifies C:

$$C_\varepsilon \text{ is defined by the function } t \to x_\varepsilon{}^A(t) \equiv x^A(t) + \varepsilon \eta^A(t), \tag{3.4}$$

where η^A are the components of an arbitrary differentiable vector function of t, which vanishes at t_a and t_b, and ε is a "small" number. Clearly, $C_{\varepsilon=0} = C$. For a fixed function $\eta^A(t)$ the extremal problem becomes a problem in extremizing I as a function of ε; then $I_\varepsilon - I_0 \propto \varepsilon^2$. In other words, if we calculate the action for the two nearby curves, one of which is the extremum curve, the first-order difference (proportional to ε) vanishes.

We now calculate the first-order terms and investigate what is implied when they vanish. Our notation is that

$$I_\varepsilon = \int_a^b L(x_\varepsilon{}^A, \dot{x}_\varepsilon{}^A, t)\, dt, \tag{3.5}$$

where the limits a, b stand for t_a, t_b and where the indication that the curve C_ε is being followed is suppressed. The vanishing of the first-order difference results in the fact that $I_\varepsilon - I_0 \propto \varepsilon^2$, where I_0 is the same as the I of (3.3). The first-order (in ε) difference is

$$
\begin{aligned}
I_\varepsilon - I_0 &= \int_a^b \left[L(x_\varepsilon{}^A, \dot{x}_\varepsilon{}^A, t) - L(x^A, \dot{x}^A, t) \right] dt \\
&= \int_a^b \left[\frac{\partial L}{\partial x^A} \varepsilon \eta^A + \frac{\partial L}{\partial \dot{x}^A} \varepsilon \dot{\eta}^A \right] dt + O(\varepsilon^2).
\end{aligned} \tag{3.6}
$$

The first-order terms are contained in the square brackets; the partial derivatives therein mean differentiation with respect to *explicit* appearances of the x^A and of the $\dot{x}^A$.

We now introduce a slight subtlety in the analysis, but one that appears in all studies of variational principles of physics. The second term in (3.6), involving $\dot{\eta}^A \equiv d\eta^A/dt$, may be rewritten by noting

$$\frac{\partial L}{\partial \dot{x}^A} \frac{d\eta^A}{dt} = \frac{d}{dt}\left(\frac{\partial L}{\partial \dot{x}^A}\eta^A\right) - \eta^A \frac{d}{dt}\frac{\partial L}{\partial \dot{x}^A}. \tag{3.7}$$

[This identity is easily verified by performing the explicit differentiation of the first term on the right in (3.7).] When (3.7) is substituted back into the integral (3.6), we recognize the integration-by-parts result:

$$I_\varepsilon - I_0 = \int_b^a \varepsilon\left(\frac{\partial L}{\partial x^A} - \frac{d}{dt}\frac{\partial L}{\partial \dot{x}^A}\right)\eta^A\, dt + \varepsilon\left(\frac{\partial L}{\partial \dot{x}^A}\eta^A\right)\Big|_{t_b}^{t_a} + O(\varepsilon^2), \tag{3.8}$$

where the term in (3.8) evaluated at the endpoint arises from the first term on the right in (3.7).

However, the vector function $\eta^A(t)$ was defined to vanish when $t = t_a$ and $t = t_b$; the endpoint terms make no contribution. Hence the first-order variation of I is given by the integral in (3.8). Recall that η^A is a completely arbitrary vector function (except for the fact that it vanishes at the endpoints) and that the integral must vanish regardless of the value of η^A. We consider such an η^A in which only one particular component, the B component, is nonzero (i.e., we consider a particular assignment B of the index A). Further, we suppose that this component has a nonzero value only for a brief interval about a given value of $t_0 \in (t_a, t_b)$. A function of a generic coordinate x, which is zero everywhere except near some particular value z_0, can be abstracted to the Dirac δ-function: $\delta(x - z_0)$, if it also satisfies $\int_{z_0-\varepsilon}^{z_0+\varepsilon} \delta(x - z_0)\, dx = 1$. In effect, we take for η^A the function

$$\eta^A = \delta_B{}^A \delta(t - t_0). \tag{3.9}$$

It then follows that the integrand of the integral in (3.8) itself vanishes at the time value t_0. Since this time is an arbitrary time within the interval in question, we simply write t for t_0; we also recognize that the index B is arbitrary and may as well be written as the general index A; thus we write for the vanishing of the integrand, after dividing by ε,

$$\frac{d}{dt}\frac{\partial L}{\partial \dot{x}^A} - \frac{\partial L}{\partial x^A} = 0. \tag{3.10}$$

This equation is equivalent to the requirement that I be an extremum under the conditions stated (cf. Exercise 2.1). The equations (3.10) are called the **Euler-Lagrange** (or simply **Lagrange**) equations. A solution of these equations is called an **actual path** for the system, whereas the other possible paths, including those in the neighborhood of an actual path, are called **potential paths** of the system (the latter are simply paths in configuration space that obey no particular equation of motion). The Lagrange equations are differential equations in time for the curve parameterized by time that extremizes the action. With a particular choice of Lagrangian function, for instance, (3.2), the connection to Newton's equations is immediate.

In the derivation of Eq. (3.10), the coordinates x^A could be Cartesian coordinates, but clearly the same derivation holds for completely general coordinates, since it is only changes in component values that enter the derivation. It is, however, useful to consider the meaning of Lagrange's equations in rectangular coordinates, in which Newton's equations read

$$m\ddot{x}^A = -V_{,A}. \tag{3.11}$$

For simple systems we have, as in Chapter 1, that the kinetic energy is a diagonal quadratic expression in the rectangular components of the velocity, and the potential in simple systems is a function only of the x^A. Hence our earlier definition of the Lagrangian $L = T - V$ reproduces Newton's equation. For multiparticle systems written in rectangular coordinates, we again verify that the Euler-Lagrange equations coincide with Newton's.

The Euler-Lagrange equation has been derived by demanding a certain geometrical object (the integral along a curve of a certain scalar function on the tangent bundle) achieve an extremum. The extremal property of a curve is independent of the coordinates used to describe the problem, and the Euler-Lagrange equations are the differential equivalent of the extremal property. One concludes then that Euler-Lagrange equations give the form, in an arbitrary coordinatization, of Newton's law of motion.

A common and useful notation is that of **functional differentiation**. The process of varying a given coordinate variable, extracting the part that is of first order in ε and then finding the integrand by, in a sense, dividing by η^A, is all very similar to finding a derivative. The notation δI is used for the variation in I. The notation $\delta x^A = \varepsilon \eta^A$ is used for the variation in the coordinate x^A. Then we have that

$$\frac{\delta I}{\delta x^A} = -\frac{d}{dt}\frac{\partial L}{\partial \dot{x}^A} + \frac{\partial L}{\partial x^A}, \tag{3.12}$$

and the Newtonian equations of motion are $\delta I / \delta x^A = 0$.

In many physical problems we want to deal with constrained systems. Holonomic constraints, as we have seen, change the dimensionality, topology, and geometry of the configuration space. Constraints of the holonomic type $f^\mu(x^A, t) = 0$ may be treated in either of two ways. The first way, which has already been discussed, makes use of the constraint equation directly to reduce the dimensionality until there are only unconstrained variables in the problem. The Lagrangian is then constructed in terms of only these unconstrained variables, and the Euler-Lagrange equations are the equations of motion. Notice that a time-dependent constraint, when eliminated, may leave a time dependence in the reduced form of a Lagrangian that originally had no explicit t dependence. One may at first wonder whether this technique gives the correct physical law of motion.

The second technique, called the **Lagrange multiplier** technique, answers this question affirmatively. The constraints are of the form $f^\mu(x^A, t) = 0$. The motion will physically be kept on the constraint surfaces by forces that act normally to the constraint surface. Such forces can be modeled by taking the constraint functions themselves with a time-varying coefficient as the potential for the forces. (We suppose that the constraint functions f^μ have nonzero gradient near $f^\mu = 0$.) The Lagrangian then is

$$\bar{L} = L - \lambda_\mu f^\mu, \tag{3.13}$$

where the λ_μ are coefficients that depend only on time, and the f^μ are the constraint functions that depend only on the coordinates and time (not velocities). In this case, the coordinates are the complete, original set x^B, $B = 1, \dots, 3N$. Notice that in this approach we use a physical model to guarantee the satisfaction of the constraints. There is no question that this scheme, if successful, correctly solves the physical problem.

The λ_μ are at this point undetermined multipliers, but they can be determined by the requirement that the constraints be satisfied. That is, the variable set $\{x^A\}$ is enlarged to the variable set $\{x^A, \lambda_\mu\}$. The Lagrangian $\bar{L}$ is used in the action integral, which is varied with respect to both x^A and λ_μ. Varying with respect to x^A yields the equations

$$\frac{d}{dt}\frac{\partial L}{\partial \dot{x}^A} - \frac{\partial L}{\partial x^A} - \lambda_\mu \frac{\partial f^\mu}{\partial x^A} = 0. \tag{3.14}$$

The result of varying with respect to λ_μ are the constraint equations

$$f^\mu = 0. \tag{3.15}$$

Equations (3.14) and (3.15) are in fact equivalent to the equations obtained by using the constraint equations to form independent variables. Exercises 3.11 and 3.12 outline the proof of this fact.

An important formal definition is that of the **canonical momentum** 1-form:

$$\mathbf{p} = p_A \, \mathbf{d}x^A, \qquad \text{where} \quad p_A \equiv \frac{\partial L}{\partial \dot{x}^A}. \tag{3.16}$$

A particular component p_A is called the momentum component **conjugate** to coordinate x^A. As a geometrical object on $\mathcal{M}$, $\mathbf{p}$ is an element of the cotangent bundle; that is, $\mathbf{p}$ is a 1-form. Because of the form of the Euler-Lagrange equations, a particular component of the canonical momentum will be constant in time (a **constant of the motion**) if the coordinate to which it is conjugate does not explicitly appear in the Lagrangian. (In such a case the coordinate is said to be **cyclic**.) The study of momenta and constants of the motion will continue in later chapters.

EXERCISES

3.1. Recall the system of Exercise 1.22. It consists of two particles P_1 and P_2. The two particles are confined to unit-radius circular tracks centered at $x = y = 0$, the first on the plane $z = 0$, the second on the plane $z = 1$. The two particles are connected by a spring with force constant k and zero natural length.

    (a) Find the Lagrangian of this system. Explain carefully what two coordinates you use and how you come to these coordinates from the Cartesian ones.

    (b) Notice that one legitimate motion of the system consists of P_1 and P_2 directly above one another and moving uniformly around their tracks. Use this observation to set up coordinates x^1 and x^2 so that L is independent of x^2.

    (c) Using the coordinates of part (b), find the general motion of the system. Is energy conserved? If it is, does the conservation of energy help?

3.2. A particle of mass m moves freely except that it is confined to the two-surface $z = x^2 + y^2$.

    (a) Using the coordinates x, y to describe the position of the particle, find the Lagrangian L (which is equal to T, the kinetic energy). Compute the canonical momenta p_x, p_y in terms of $\dot{x}, \dot{y}$. Explain in detail how to solve for the motion in terms of the coordinates x, y. Do not perform the actual calculation.

    (b) Repeat the calculation by using all three coordinates x, y, z and using a Lagrange multiplier. Show that equivalent equations of motion are obtained.

3.3. Write the motion of a simple pendulum using plane-polar coordinates r, θ (where $r = 0$ is the support point) using the method of Lagrange multipliers to impose the constraint $r = \ell = \text{const}$. Write out the resulting Lagrangian and the equations of motion, and finally reduce the equations so that there is a single equation in which only one variable appears, and so that the Lagrange multiplier is explicitly determinable. Then assume small amplitude oscillations and give the complete solution for the motion and for the Lagrange multiplier.

3.4. Consider a point particle of mass m, which moves under the influence of gravity while simultaneously constrained to lie on a sphere. Write the Lagrangian in spherical coordinates, including the constraint, via a Lagrange multiplier. Then determine an expression for the Lagrange multiplier and eliminate it from the equations. Reduce the equations (do not solve them) to a form where you can discuss the physical meaning of the different terms that appear. Carry out this discussion.

3.5. A **metric tensor** $g_{\alpha\beta}$ is a symmetric tensor whose array of components is a nonsingular matrix function of position in any coordinate system $\{x^A\}$. It is used to define distance. Consider motion restricted to a plane. The distance along an arbitrary curve between two points A, B is

$$\text{distance} = \int_B^A ds,$$

where

$$ds = \sqrt{ds^2} = \sqrt{g_{\alpha\beta}\, dx^\alpha\, dx^\beta} = \sqrt{g_{\alpha\beta}\dot{x}^\alpha\dot{x}^\beta}\, d\lambda,$$

where λ is an arbitrary parameter along the curve and $\dot{}$ is $d/d\lambda$. Use plane-polar coordinates. By considering different possible curves, show that the shortest distance between two points is a straight line. That is, verify that for fixed endpoints the variation of the distance integral with respect to each of the two polar coordinates yields equations that are equivalent to the equation for a straight line. Try the calculation in general, but then consider a hint, which makes this problem more tractable. Because λ is arbitrary, it can be redefined at any point to simplify the problem. In particular, one can switch to an especially useful parameter $\bar{\lambda}$ defined by $d\bar{\lambda} = ds$; this transformation is a legitimate one because every variable is parameterized by λ.

Further, show that the integral (which is simpler than the one above because it is without the square root)

$$\int_A^B g_{\alpha\beta}\dot{x}^\alpha\dot{x}^\beta\, d\lambda$$

gives the same straight line as the first integral does, although this integral in addition requires that λ be used as the path parameter. A path that is an extremum path for this integral is called a **geodesic**, and the path parameter is called an **affine parameter**.

3.6. In the theory of relativity, space and time coordinates can mix under Lorentz transformations, which describe rectangular frames moving relatively at fixed velocity. There is an invariant of interest, a generalization of the Euclidean distance element:

$$ds^2 = -dt^2 + dx^2 + dy^2 + dz^2.$$

Under Lorentz transformations, this quantity has the same form and same value whether computed in one or the other coordinate system. In this four-dimensional spacetime, photons follow orbits $x^\alpha(\lambda)$, $\alpha = 0, 1, 2, 3$ with λ a parameter, where $x^0 = t$, $x^1 = x$, $x^2 = y$, $x^3 = z$. The light paths are determined by the requirement that they be **null geodesics**; that is, they not only obey the geodesic condition of the preceding problem but also obey the condition $ds^2 = 0$. More explicitly, the integral to be used as the action integral is

$$I = \int \left(-\dot{t}^2 + \dot{x}^2 + \dot{y}^2 + \dot{z}^2 \right) d\lambda,$$

and the null condition is

$$-\dot{t}^2 + \dot{x}^2 + \dot{y}^2 + \dot{z}^2 = 0.$$

Find the momentum conjugate to the variables x^α. Show that they are constants of the motion. Find the null geodesic orbits that the photons follow.

3.7. Consider a **cosmological model** in which the metric is given by

$$ds^2 = -dt^2 + R^2(t)\left[dr^2 + \sin^2 r(d\theta^2 + \sin^2\theta\, d\phi^2)\right],$$

where R is a given function of t, which is monotonic increasing and which vanishes at $t = 0$. Although the origin $r = 0$ appears special in this formulation, this is actually a homogeneous model and any point can be chosen as the origin. Use this metric in the same way the metric was used in the preceding exercise to find the explicit action integral, which, when extremized, yields the equations geodesics must obey. Find the further condition that characterizes null geodesics.

(a) Find all radial lines (only t and r varying) that are null geodesics.

(b) Consider this model universe at some $t = t_0$. From what radius in r come those radial null geodesics that reach the origin $r = 0$ at $t = t_0$ and that started at $t = 0$? Since the model universe presumably began at $t = 0$ [since $R(0) = 0$], this radius is a **horizon** beyond which no information can come.

(c) In this model, $r = \pi$ is the "antipode" of the origin, the opposite pole of a three-sphere space on which $\{r, \theta, \phi\}$ are used as coordinates. Suppose that $R = R_0(t/t_0)^p$, where t_0 and p are constants. For what value of p (and t) can one see around the universe?

3.8. Let $L(x^i, \dot{x}^i, t)$ be the Lagrangian of a system. Let $\bar{L} = L + dF/dt$, where $F = F(x^i, t)$, be the Lagrangian of a second system in the same configuration space as the first. Show that if $x^i(t)$ is an actual path for one system, $x^i(t)$ is also an actual path for the other. Let $M(x^1, \dot{x}^i, t)$ be still another Lagrangian for another system having the same configuration space. Suppose that M is a function of L, with $dM/dL \neq 0$. Under what conditions do both Lagrangians yield the same actual paths?

3.9. Suppose that we have a system with a Lagrangian without constraints, but where

$$L = L(x^i, \dot{x}^i, \ddot{x}^i, t).$$

Obtain the equations of motion (Euler-Lagrange equations) in this case. Assume that all the derivatives of the variation vanish at the endpoints. Discuss whether such a system could be a good representation for a mechanical system.

3.10. Two masses m_1, m_2, which lie on a frictionless surface, are connected by a spring (with spring constant k_2 and zero natural length), and m_1 is connected to a fixed (immovable) wall by a spring (spring constant k_1 and zero natural length).

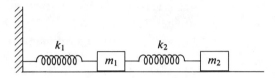

Figure P3.10

(a) What is the dimension of the configuration space $\mathcal{M}$? Define some useful coordinates to label points in $\mathcal{M}$. Discuss the topology of $\mathcal{M}$. (For example, is it a line, a plane, a sphere, or part of one of these?)

(b) Write down the Lagrangian and the Euler-Lagrange equations for this system.

(c) Obtain explicitly the canonical momenta of this system. Is energy conserved? What are the coordinates and momenta physically?

3.11. Suppose that a Lagrangian $L(x^A, \dot{x}^A, t)$ is given in Cartesian coordinates but that these coordinates are actually functions of the independent coordinates q^i and time t: $x^A = x^A(q, t)$. Because of these relations, the velocity components are given by

$$ \dot{x}^A = \frac{\partial x^A}{\partial q^i}\, \dot{q}^i + \frac{\partial x^A}{\partial t}. $$

Define the Lagrangian $\bar{L}(q^i, \dot{q}^i, t)$ by

$$ \bar{L}(q^i, \dot{q}^i, t) = L\left(x^A(q,t),\, \frac{\partial x^{A\,i}}{\partial q}\, \dot{q}^i + \frac{\partial x^A}{\partial t}, t\right). $$

This is the Lagrangian that results when the constraint equations are used to express the Cartesian coordinates in terms of independent coordinates. Show that if the Euler-Lagrange equations for L are satisfied (for variations in the x^A), then the Euler-Lagrange equations for $\bar{L}$ (for variations in the q^i) will also be satisfied: Compute the Euler-Lagrange equations for $\bar{L}$ and show that they are the Euler-Lagrange equations for L times the partial derivatives $\partial x^A / \partial q^i$.

3.12. In Exercise 3.11 the Euler-Lagrange equations for L may not in fact be satisfied. Suppose instead that Eqs. (3.14) are satisfied. The constraint equations $f^\mu(x, t) = 0$ are postulated to be independent and equivalent to the relations $x^A = x^A(q, t)$. Show that under these conditions the Euler-Lagrange equations for $\bar{L}$ are satisfied. This demonstrates that Eqs. (3.14) and (3.15) are equivalent to the equations obtained by using the constraint equations to form independent variables.

4

CENTRAL FORCE FIELDS

In this chapter we apply the techniques described in Chapter 3 to the study of the motion of two particles under the influence of a force that depends only on the distance between them. This application is important for celestial mechanics and in classical analogs of systems such as the hydrogen atom where the effects of quantum mechanics dominate. An understanding of classical scattering theory also introduces the concepts essential for understanding and interpreting quantum mechanical scattering.

In fact, we start with the N-body case and set up the Lagrangian and equations of motion in convenient coordinates. We then study the orbits for the case of two bodies and describe the important conserved quantities. Solutions expressible as analytic combinations of simple functions are in general not possible for general central forces, but for the Coulomb force—that is, a force that varies inversely as the square of the separation distance—such solutions can be found. We discuss them and also modifications of the Coulomb force. The chapter ends with discussions on scattering theory and three- and N-body celestial mechanical systems.

4.1 General Considerations

Consider a general case: N particles, with no constraints, such that the potential function V depends only on the relative positions of the particles. In terms of the Cartesian coordinates of the particles x^A, the Lagrangian $L = T - V$ is

$$L = T_{AB}\, u^A u^B - V, \tag{4.1}$$

where u^A is the velocity, $u^A = \dot{x}^A = dx^A/dt$. The special functional form of V allows a simplification. First, we change our notation somewhat, returning to a notation that emphasizes the individuality of the particles. Let m_a be the mass of particle a (where $a = 1, \ldots, N$), and let $x^i{}_a$ be the ith coordinate (here $i = 1, 2, 3$) of particle a. The kinetic energy term T is then

$$T = \tfrac{1}{2} \sum_{a=1}^{N} m_a\, \delta_{ij}\, u^i{}_a u^j{}_a. \tag{4.2}$$

Notice that we use the summation convention for the dummy indices i, j (each takes the range $1, 2, 3$) but put in an explicit summation sign for the particle number index a, since the notation does not obey the summation convention.

The coordinates of the center of mass X^i of this system are defined by

$$X^i = \frac{1}{M} \sum_{a=1}^{N} m_a\, x^i{}_a, \tag{4.3}$$

where M is the total mass:

$$M = \sum_{a=1}^{N} m_a. \tag{4.4}$$

Equation (4.3) defines the center of mass as the mass-weighted average of particle positions.

Let $\xi^i{}_a$ be the position coordinates of particle a relative to X^i:

$$\xi^i{}_a \equiv x^i{}_a - X^i. \tag{4.5}$$

Then V depends only on the various differences $\xi^i{}_a - \xi^i{}_b$, since we had previously specified that V depends only on the differences in the x^i coordinates. We leave as an exercise (see Exercise 4.2) the proof that

$$\sum_{a=1}^{N} m_a\, \xi^i{}_a = 0. \tag{4.6}$$

We also leave as an exercise (Exercise 4.2) the proof that

$$T = \frac{1}{2} M \delta_{ij}\, U^i U^j + \frac{1}{2} \sum_{a=1}^{N} m_a\, \delta_{ij}\, w^i{}_a w^i{}_a, \tag{4.7}$$

where $U^i = \dot{X}^i = dX^i/dt$ and $w^i{}_a = \dot{\xi}^i{}_a = d\xi^i{}_a/dt$. Due to the relations (4.6), it is clear that the $\xi^i{}_a$ comprise a set of only $3N - 3$ independent coordinates. We may define the three last of them in terms of the others:

$$\xi^i{}_N = -\frac{1}{m_N} \sum_{a=1}^{N-1} m_a \xi^i{}_a. \tag{4.8}$$

We now take as the $3N$ coordinates of our system $\{X^i, \xi^i{}_a\}$ (for $a = 1, \ldots, N - 1$). The potential energy term V is a function only of these $\xi^i{}_a$ and is independent of the center of mass coordinates X^i. Further, from (4.7) we see that T is of the form

$$T = \tfrac{1}{2} M \delta_{ij} U^i U^j + \bar{T}, \tag{4.9}$$

where $\bar{T}$ is independent of X^i and U^i. Notice also that $\bar{T}$ is independent of the coordinates $\xi^i{}_a$, being homogeneous quadratic in the velocities $w^i{}_a$ ($a = 1, \ldots, N - 1$).

The equations of motion are, as always, the Euler-Lagrange equations for $L = T - V$. Three of these are

$$\frac{d}{dt}\frac{\partial L}{\partial U^k} = 0, \tag{4.10}$$

since L is independent of X^k (i.e., X^k is cyclic). These equations are particularly simple: The acceleration of the center of mass vanishes. Therefore, the center of mass moves at a constant velocity:

$$X^k = U_0^k t + X_0^k, \tag{4.11}$$

where U_0^k is the constant velocity of the center of mass and X_0^k is the set of constants describing the position of the center of mass at $t = 0$.

Here is an example where a partial solution may be represented by a set of constraints. We define three constraint functions f^k ($k = 1, 2, 3$) by

$$f^k = X^k - U_0^k t - X_0^k. \tag{4.12}$$

The constraint requirements

$$f^k = 0, \tag{4.13}$$

first of all, represent merely what the X^k coordinates would be doing. On the other hand, the imposition of $f^k = 0$ as constraints allows us to simplify the problem by taking as configuration space **C** the $(3N - 3)$-dimensional space with coordinates $\xi^i{}_a$ ($k = 1, 2, 3$; $a = 1, \ldots, N - 1$). Further reduction of the general problem in this manner is possible, but for now we restrict ourselves to the case of just two particles.

4.2   The Two-Body Problem

When $N = 2$, the foregoing method shows that the Lagrangian L may be reduced to the form $\bar{T} - V$ and that configuration space $\mathbf{C}$ is three-dimensional. As an exercise, show that $L = \bar{T} - V$ becomes

$$L = \tfrac{1}{2}\mu\,\delta_{ij}\,v^i v^j - V, \tag{4.14}$$

where μ is the reduced mass, defined by

$$\mu = \frac{m_1 m_2}{m_1 + m_2}, \tag{4.15}$$

and where the velocity compoments v^i are the time derivatives $v^i = \dot{q}^i$ of coordinates q^i defined by

$$q^i = \frac{1}{\mu}\,m_1\,\xi^i{}_1 = -\frac{1}{\mu}\,m_2\,\xi^i{}_2. \tag{4.16}$$

The potential energy function V is expressible as a function of q^i, of course [see Exercise 4.3(a)].

The reduction described above is equivalent to choosing coordinates in which the center of mass lies at the origin. This choice is always possible for an isolated system obeying Newton's third law, as we will emphasize later. In terms of the q^k coordinates, the coordinates $x^i{}_a$ of the individual particles are obtained by solving (4.16), since $x^i{}_a = \xi^i{}_a$ because $X^i = 0$.

In general, V will depend on all three of the coordinates q^i, but we now specialize to the case of a potential dependent only on the distance between the two particles. Thus V is a function only of this distance r, defined by

$$r = \left[\delta_{ij}(x^i{}_1 - x^i{}_2)\,(x^j{}_1 - x^j{}_2)\right]^{1/2}. \tag{4.17}$$

Notice that in terms of q^k, r is given by

$$r = [\delta_{ij}\,q^i q^j]^{1/2}. \tag{4.18}$$

We therefore can adopt the usual spherical coordinates $\{r, \theta, \phi\}$ from which the q^k are given by

$$q^1 = r\sin\theta\cos\phi, \qquad q^2 = r\sin\theta\sin\phi, \qquad q^3 = r\cos\theta. \tag{4.19}$$

In $\{r, \theta, \phi\}$ coordinates, L is

$$L = \tfrac{1}{2}\mu[\dot{r}^2 + r^2\dot{\theta}^2 + r^2\sin^2\theta\,\dot{\phi}^2] - V(r). \tag{4.20}$$

The equations of motion, the Euler-Lagrange equations, are (˙ means d/dt, as always):

$$\left(\mu r^2 \sin^2\theta\,\dot\phi\right)^{\cdot} = 0,$$

$$\left(\mu r^2\dot\theta\right)^{\cdot} - \mu r^2 \sin\theta\cos\theta\,\dot\phi^2 = 0, \tag{4.21}$$

$$\left(\mu\dot r\right)^{\cdot} - \mu r\dot\theta^2 - \mu r\sin^2\theta\,\dot\phi^2 + \frac{dV}{dr} = 0.$$

The equations for θ and ϕ are particularly simple, since V is independent of θ and ϕ.

We can simplify these equations further by making use of the conservation of angular momentum ℓ_k. The total angular momentum of a system of N particles with respect to the origin is

$$\ell_k = \sum_{a=1}^{N} m_a\,\varepsilon_{kij}\,x^i{}_a u^j{}_a, \tag{4.22}$$

where ε_{kij} is the Levi-Civita tensor, the completely antisymmetric third-rank tensor defined by requiring $\varepsilon_{123} = 1$. For our system of two particles, ℓ_k takes the form

$$\ell_k = \mu\,\varepsilon_{kij}\,q^i v^j, \tag{4.23}$$

so that the masses appear only in the reduced mass μ.

To show that ℓ_k is conserved, we first go back to the expression of the Lagrangian in q^k coordinates. In dealing with V as a function of q^k, because of its dependence on r we note that

$$\frac{\partial V}{\partial q^k} = \frac{dV}{dr}\frac{\partial r}{\partial q^k} = V'\frac{1}{r}q^\ell\delta_{k\ell}, \tag{4.24}$$

where $V' \equiv dV/dr$. The Euler-Lagrange equations in these coordinates are

$$\mu\,\delta_{k\ell}\,\ddot q^\ell + \frac{1}{r}V'\delta_{k\ell}\,q^\ell = 0. \tag{4.25}$$

We left the $\delta_{k\ell}$ in explicitly, since we want to keep to our convention that q^k be written only with a superscript index, but clearly this equation can be then written in the simpler form

$$\ddot q^\ell = -\frac{V'}{\mu r}\,q^\ell. \tag{4.26}$$

To solve the equations in this form, we must remember that r is a function of q^k.

The time derivative of the angular momentum ℓ_k is

$$\dot\ell_k = \mu\,\varepsilon_{k\ell m}(v^\ell v^m + q^\ell\ddot q^m). \tag{4.27}$$

The first term vanishes because $\varepsilon_{k\ell m}$ is antisymmetric in ℓ, m, but $v^\ell v^m$ is symmetric. The second term vanishes because, via (4.26), $\ddot q^m$ is proportional to q^m, and we can again invoke the statement that an antisymmetric tensor double-summed with a symmetric one gives zero. Hence $\dot{\ell}_k = 0$.

Since ℓ_k is constant, we may rotate our q^k system to produce new coordinates q^k in which ℓ_k has only a component in the q^3-direction. We leave it as an exercise [see Exercise 4.3(b)] to show that if r, θ, ϕ were defined in terms of the q^k coordinates, then $\theta = \pi/2 = \text{const.}$ is a consequence of the equations of motion. Here is another example where a partial solution of the problem (i.e., the use of the conservation of angular momentum) is equivalent to choosing a constraint, namely, $\theta - (\pi/2) = 0$. Like the restriction to center-of-mass behavior, this is a natural separation of the variables of the problem, and, in particular, no constraint force is required to satisfy this constraint.

At any rate, with this constraint the system's configuration space becomes the two-dimensional plane described by polar coordinates $\{r, \phi\}$. The Lagrangian is

$$L = \tfrac{1}{2}\mu[\dot{r}^2 + r^2\dot{\phi}^2] - V(r), \tag{4.28}$$

and the equations of motion are

$$[\mu r^2 \dot{\phi}]\dot{} = 0, \tag{4.29}$$

$$\mu\ddot{r} - \mu r\dot{\phi}^2 + V' = 0. \tag{4.30}$$

The first equation has a partial solution:

$$\mu r^2 \dot{\phi} = \ell = \text{const.}, \tag{4.31}$$

where ℓ is the magnitude of the angular momentum vector ℓ_k. This equation can be solved for ϕ only when the time dependence of r is determined.

The equation for r now can be put into the form

$$\mu\ddot{r} - \frac{\ell^2}{\mu r^3} + V' = 0. \tag{4.32}$$

This is now a one-dimensional problem; a standard trick for integrating an equation of this form (where t does not explicitly appear) is to use r as the independent variable. Then write, using $s \equiv \dot{r}$,

$$\ddot{r} = \frac{ds}{dr}\frac{dr}{dt} = s\frac{ds}{dr} = \frac{1}{2}\frac{d(s^2)}{dr}. \tag{4.33}$$

Equation (4.32) is then easily integrated once to yield

$$\frac{1}{2}\mu s^2 + \frac{\ell^2}{2\mu r^2} + V = E = \text{const.} \tag{4.34}$$

Alternatively, one may multiply (4.32) by $\dot{r}$ and recognize $2\dot{r}\ddot{r} = (\dot{r}^2)\dot{}$, achieving (4.34) again. Further integration requires knowledge of $V(r)$.

We have now reached a point where only two equations, (4.31) and (4.34), must be solved to find the **trajectory** (the parameterized path), namely, $(r(t), \phi(t))$. The **orbit** is the equation for r as a function of ϕ. The orbit tells the subset of configuration space traveled by the system, but it cannot give the velocity (for which the actual trajectory is needed). To find the equation for the orbit, we write r as a function of ϕ alone, implying that $s = (dr/d\phi)\,(d\phi/dt)$. Since (4.31) gives $\dot{\phi}$ as a function of r, we find that (4.34) becomes

$$\frac{1}{r^4}\left(\frac{dr}{d\phi}\right)^2 + \frac{1}{r^2} + \frac{2\mu}{\ell^2}\,(V - E) = 0, \qquad (4.35)$$

(assuming that $\ell \neq 0$). If we let $u = 1/r$, this equation takes the even simpler form

$$\left(\frac{du}{d\phi}\right)^2 + u^2 + \frac{2\mu}{\ell^2}\,(V - E) = 0. \qquad (4.36)$$

In either case, the equation can be solved for by an immediate integral (i.e., solved by quadrature), as we show in the next section.

One immediate deduction can be made even at this general stage before a specific form of V is chosen: From (4.34) or (4.36) we see that $V - E$ must be negative, so that the values of u for which $V - E$ is positive are not allowed. If V increases indefinitely with increasing u, that is, with decreasing r, then u is bounded from above: Then there is a minimum value of r_0 below which r cannot go.

4.3   The Orbit Equation

We have now reduced the six-dimensional configuration space to a two-dimensional space by making use of the conservation of total linear momentum and total angular momentum. In both cases the conservation laws represented partial solutions to the equations of motion. In both cases these solutions could be used in the form of constraints to reduce the dimensionality of configuration space. In short, these solutions allowed us to choose as configuration space a plane parameterized by r and an angle ϕ, the constraints being that the center of mass is centered at the origin and the angular momentum is oriented to point along the x^3-axis.

The potential function V depends on r but not on ϕ. The Lagrangian L is

$$L = \tfrac{1}{2}\mu[\dot{r}^2 + r^2\dot{\phi}^2] - V(r). \qquad (4.37)$$

Remember that μ is the reduced mass, formed from the two masses m_1 and m_2:

$$\mu = \frac{m_1 m_2}{M} \qquad \text{with} \quad M = m_1 + m_2. \qquad (4.38)$$

The coordinates q^k ($k = 1, 2, 3$; but $q^3 = 0$) are determined from r, ϕ, and then the positions of the masses $x^k{}_1$ and $x^k{}_2$ are determined by

$$q^1 = \frac{M}{m_2}\,x^1{}_1 = -\frac{M}{m_1}\,x^1{}_2 = r\cos\phi,$$

$$q^2 = \frac{M}{m_2} x^2{}_1 = -\frac{M}{m_1} x^2{}_2 = r \sin \phi, \qquad (4.39)$$

$$q^3 = \frac{M}{m_2} x^3{}_1 = -\frac{M}{m_1} x^3{}_2 = 0.$$

The equations of motion are Eqs. (4.29)–(4.31). The constant ℓ in (4.31) is the magnitude of the angular momentum. The procedure is then to use this result in (4.30) for $r(t)$, and once $r(t)$ is determined, solve for $\phi(t)$. Equation (4.31) results in **Kepler's second law** of planetary motion: The line joining a mass to the center of mass sweeps out equal areas during equal time time intervals in its orbit provided that the force depends only on the distance between the two masses. Notice that the functional form of $V(r)$ does not matter. [Equation (4.31) holds even if V is a function of t as well as r.]

The term involving ℓ^2 in Eq. (4.34) is one whose negative r-derivative is centrifugal force. It may be denoted

$$V_c = \frac{\ell^2}{2\mu r^2}, \qquad (4.40)$$

and it is called the centrifugal barrier because of its role in preventing r from becoming zero. The total of V_c and V is an effective potential

$$V_{\text{eff}} \equiv V_c + V. \qquad (4.41)$$

The formal solution to the equation for r is obtained by first solving (4.34) for t:

$$t - t_0 = \int_{r_0}^{r} \left[\frac{2}{\mu} (E - V - V_c) \right]^{-1/2} dr, \qquad (4.42)$$

where $r = r_0$ at $t = t_0$. This result is formal in that it is only in principle that this equation may be inverted to find $r(t)$. When $r(t)$ is found, $\phi(t)$ is, again in principle, obtainable as a well-defined integral via Eq. (4.29). Although these integrals are typically not expressible in terms of tabulated functions, this form of the solution is certainly useful. The reduction of a problem to the point where mere one-variable integrals remain is called reducing the problem to *quadratures*.

The equation for the orbit $r(\phi)$ is obtained by writing time derivatives as

$$\frac{d}{dt} = \frac{d\phi}{dt} \frac{d}{d\phi} = \frac{\ell}{\mu r^2} \frac{d}{d\phi}. \qquad (4.43)$$

It is best to work with the second-order equation (4.32). First write $\ddot{r}$ as

$$\ddot{r} = \frac{\ell}{\mu r^2} \frac{ds}{d\phi}, \quad \text{with} \quad s = \dot{r}. \qquad (4.44)$$

Next use

$$u \equiv \frac{1}{r}, \quad \text{so that} \quad \frac{du}{d\phi} = -\frac{1}{r^2} \frac{dr}{d\phi}. \qquad (4.45)$$

The equation for $\ddot{r}$ is further simplified when it is recognized that V itself may be expressed as a function of u rather than r. We have

$$\frac{dV}{dr} = \frac{dV}{du}\frac{du}{dr} = -u^2\frac{dV}{du}. \tag{4.46}$$

The equation for $\ddot{r}$, when written as an equation in $u(\phi)$, is then

$$\frac{d^2u}{d\phi^2} + u + \frac{\mu}{\ell^2}\frac{dV}{du} = 0. \tag{4.47}$$

The beauty of (4.47) is apparent for certain force laws. For instance, if $V = 0$, no force at all, the equation is

$$\frac{d^2u}{d\phi^2} + u = 0. \tag{4.48}$$

This equation is the simple harmonic oscillator equation (but what a different physical situation!) and has the solution

$$u = \frac{1}{r} = u_0\cos(\phi - \phi_0). \tag{4.49}$$

The constants u_0 and ϕ_0 are determined by "initial" data, that is, data at any point in the orbit where position and velocity may be specified.

It is easily seen that (4.49) is the equation of a straight line. The value $u = u_0$ corresponds to $\phi = \phi_0$; it is the maximum value possible for u; the minimum value for r is $r = r_0 = 1/u_0$, the distance of closest approach. We note that $u_0 > 0$: Unlike the simple harmonic oscillator situation, here $u < 0$ has no meaning. The value $u = 0$ corresponds to $r = \infty$, which occurs at $\phi = \phi_0 \pm \pi/2$. The whole angle between $r = \infty$ on one side and $r = \infty$ on the other side is $\Delta\phi = \pi$.

4.4 The Coulomb Force

An inverse square force law is generated by a potential function V proportional to r^{-1}:

$$V = \frac{k}{r} = ku. \tag{4.50}$$

Positive k corresponds to a repulsive force with components $f_i = -V_{,i}$. (Negative values of the constant k are certainly allowed; they correspond to attractive forces.) The equation for $u(\phi)$ is then

$$\frac{d^2u}{d\phi^2} + u + \frac{k\mu}{\ell^2} = 0. \tag{4.51}$$

The solution of this equation is the sum of the general solution to the homogeneous ($k = 0$) equation and the particular solution $u = \text{const}.$:

$$u = u_0 \cos(\phi - \phi_0) - \frac{k\mu}{\ell^2}. \tag{4.52}$$

Because of the periodicity of the cosine, ϕ_0 may be chosen in a way that makes $u_0 > 0$. Then the angle $\phi = \phi_0$ corresponds to the maximum value of u, that is, to the distance of closest approach. Notice, however, that $u_{max} \neq u_0$ (unless $k = 0$). For convenience in our discussion below, we simply choose $\{r, \phi\}$ coordinates so that $\phi_0 = 0$, with $u_0 > 0$.

Consider the attractive case ($k < 0$) first. The solution is of the form

$$u = \frac{1}{r} = \frac{\mu |k|}{\ell^2}(1 + \varepsilon \cos \phi), \tag{4.53}$$

where

$$\varepsilon = \frac{u_0 \ell^2}{\mu |k|}. \tag{4.54}$$

Note that the value of ε is by definition nonnegative. If $\varepsilon = 0$, the orbit is circular: $r = \text{const}$. If $0 < \varepsilon < 1$, u is positive for all values of ϕ. The orbit is thus closed or periodic in the sense that it is a closed figure, and the motion continually repeats itself. The orbit may also be expressed in $\{q^1, q^2\}$ coordinates (remember that $q^3 = 0$) using (4.39). Then the equation for the orbit (4.53) may be put into the form

$$(q^2)^2 + (1 - \varepsilon^2)(q^1)^2 + \frac{2\varepsilon \ell^2}{\mu |k|} q^1 = \frac{\ell^4}{\mu^2 k^2}. \tag{4.55}$$

Clearly when $0 < \varepsilon < 1$, this is an ellipse. When $\varepsilon = 1$, the coordinate q^1 appears only linearly in Eq. (4.55), which is then the equation of a parabola. By referring to the equation for u, it is seen that the ε case has $u \to 0$ for $\phi \to \pm\pi$. This parabolic orbit has the particle coming in from an infinite distance and returning there in the same direction. If $\varepsilon > 1$, $u \to 0$ for values of ϕ bounded away from $\pm\pi$. This case is of an open or hyperbolic orbit.

In the repulsive force ($k > 0$) case, u will be negative (negative values being, of course, not allowable), unless

$$u_0 > \frac{\mu k}{\ell^2}. \tag{4.56}$$

The orbits are always hyperbolic: $u \to 0$ for values of ϕ bounded away from $\phi = \pm\pi$.

We have so far not addressed the case of the actual velocity components, determined by $\dot{r}$ and $\dot{\phi}$. Rather than treat the velocity in detail, we will consider the velocity component $\dot{r}$ only. Remember that $\dot{\phi} = \ell/\mu r^2$, so that

$$\dot{r} = \frac{d\phi}{dt}\frac{dr}{d\phi} = \frac{\ell}{\mu} u_0 \sin \phi. \tag{4.57}$$

Consequently, $\dot{r} = 0$ whenever $\phi = 0$ (closest approach), or whenever $\phi = \pm\pi$. In the closed or parabolic orbits of the $k < 0$ case, ϕ does reach $\pm\pi$, and $\dot{r} \to 0$

then. In the parabolic case, in other words, the particle reaches infinite distance at asymptotically zero speed (since $\dot{\phi} \to 0$ as $r \to \infty$). In the hyperbolic case, there is residual nonzero velocity as $r \to \infty$, for the angle ϕ never achieves the values $\pm\pi$.

Not all force laws lead to exact closed analytic solutions in terms of simple functions. One that does with the techniques we have been using leads to somewhat surprising new properties for the orbit. Suppose that the force is the sum of an attractive k/r^2 term and a term of the form B/r^3 (note that positive B corresponds to a repulsive addition to the inverse square force). The equation for u then becomes

$$\frac{d^2u}{d\phi^2} + u = \frac{\mu|k|}{\ell^2} - \frac{\mu B}{\ell^2} u. \qquad (4.58)$$

The inverse cube force term thus mimics a centrifugal force term. From (4.58) it is clear that it modifies the period of u as a function of ϕ. Let ω be defined by

$$\omega \equiv \left(1 + \frac{\mu B}{\ell^2}\right)^{1/2}. \qquad (4.59)$$

Then the solution for u is

$$u = \frac{\mu|k|\omega}{\ell^2}\left(1 + \varepsilon\cos\omega\phi\right). \qquad (4.60)$$

If the B term is attractive ($B < 0$), ϕ must increase by more than 2π to cause u to return to the same value. If B is small, the orbit remains qualitatively an ellipse, but the ellipse precesses. The position of the perihelion (position of closest approach to the sun) of a planet that obeys this kind of force law advances per period by the angle $\Delta\phi$ given by

$$\Delta\phi = 2\pi(\omega - 1) \approx \frac{\pi\mu B}{\ell^2}. \qquad (4.61)$$

If B is negative, the precession still obeys this law, but the orbit precesses backward.

Note that for a near circular orbit and small B, $\ell^2 \simeq \mu|k|a$ where a is the orbital radius. Hence

$$\Delta\phi \simeq \pi\frac{B}{|k|a}. \qquad (4.62)$$

4.5   Classical Scattering Theory

The gravitational potential of a macroscopic object such as the sun or a planet may be deduced from the behavior of the orbits of satellites. The interaction between a microscopic body and other particles is much less accessible to observation. The common method is to bombard a target particle with a monoenergetic, directed stream of particles and to observe the deflections from straight-line motion of their (unbound) orbits. The result is expressed as a **cross section**, having the units of an area, for the scattering particle. Of course, the idea of a scattering cross section can be applied to macroscopic targets such as the sun, even though it is not usual to do so. Too, a correct treatment of the microscopic case requires quantum mechanics. Nevertheless, it is useful to treat classical scattering in much the same language as is used in cases requiring quantum mechanics, and that is the subject of this section. In the important case of scattering by the Coulomb (inverse square) force, the classical and quantum results (for spinless particles) agree remarkably. In our analysis we will assume that the potential is *central;* describing the interaction depends only on r, the distance between the target and bombarding particles.

In a scattering experiment, the energy of the bombarding particles is typically much more than the binding due to any forces (chemical forces) holding the target particle in place. Thus the target, which we assume is placed at the origin, is free to recoil. Nevertheless, we conduct the analysis by assuming that the target is so massive that the recoil is negligible. As we have seen, any two-body problem can be recast into a single-particle form by introducing relative coordinates and the reduced mass. We will then still have to include effects of recoil, and we will demonstrate that at the end of the chapter.

The experiment consists of projecting a beam of particles toward the target. The particles in the beam initially move parallel to each other, and each has the same initial speed v_0 (equivalently, we could say that the particles each have the same initial kinetic energy). The particles all have the same mass m. As we said above, the target is placed at the origin; the beam is set to impinge on the target parallel to the ($z = x^3$)-axis, from $z = -\infty$. The initial distance of a given particle in the beam from the z-axis is called its **impact parameter** b:

$$b = \left(x^2 + y^2\right)_{\text{initial}}^{1/2} . \tag{4.63}$$

After scattering, the particles emerge in various directions, but the orbit of any given particle always lies in the plane containing its initial velocity and the z-axis. The target actually consists of many particles, so spread out that each bombarding particle (which we will call a *bullet*) can be assumed to interact with but one target particle, and each *target* interacts with at most one bombarding bullet.

Nonetheless, the scattering takes place at a level where individual target particles cannot be seen. Thus we will analyze a typical target-bullet pair but then will apply statistical or averaging techniques to the result. These techniques will use $d^2N/dA\,dt$, the number of particles per unit area per unit time of the bombarding beam; we assume that this **beam density** is constant. If we consider an area dA (perpendicular to the z-axis and far below the origin) and count the number dN of particles crossing this area in time dt, then

$$dN = \frac{d^2N}{dA\,dt}\,dA\,dt. \tag{4.64}$$

We assume that the orbit of each bullet-target pair is adequately described by the two-body equations developed previously in this chapter. This assumption is the same as the assumption that the scattering is elastic: The initial total kinetic energy equals the final total kinetic energy. (We will not discuss the important case that often happens in real experiments, when either target or bullet may absorb or emit energy, causing an inelastic collision.) In the elastic case, because the target is initially at rest, the total energy E, which is conserved, is equal to the initial kinetic energy of the bullet so that at all times

$$E = \tfrac{1}{2}mv_0^2 = \tfrac{1}{2}m\dot{z}_{\text{initial}}^2 . \tag{4.65}$$

Conservation of angular momentum is the reason that the scattering of each bullet is in a plane. In addition, the magnitude of angular momentum ℓ at any time is the same as its initial value,

$$\ell = mv_0 b. \tag{4.66}$$

For clarity we will be assuming a repulsive central force; the attractive case is very similar. We will also assume that $|V(r)|$ decreases monotonically to zero as $r \to 0$.

Consider all bullets that have impact parameters between b and $b + db$. The area of this annulus is $2\pi b\, db$, and the total number of the bullets that pass through it per unit time is $2\pi b\, db (d^2 N/dA\, dt)$. Because of the repulsive force, those particles initially at distance b from the z-axis are deflected by an angle $\Theta \geq 0$ away from the z-axis. Those at distance $b+db$ are deflected by a smaller angle $\Theta + d\Theta$, $d\Theta \leq 0$ (because they stay farther away from the force center). Hence the solid angle into which the particles scatter is $d\Omega = 2\pi |d\Theta| \sin \Theta$. Thus $d\sigma = 2\pi b\, db$ is the **cross section** (area) of the incident beam, which scatters into the solid angle $d\Omega$ associated with the **scattering angle** Θ. Note that Θ is a function of b and may be determined from the orbit equation appropriate to the potential being used.

When microscopic objects do the scattering, individual bullets do not tell much information because their impact parameters cannot be known. In this case it is sensible to talk about the solid angle of the detector. One usually is interested in the **differential cross section** $d\sigma/d\Omega$ for scattering per unit solid angle. This quantity is the area of the incident beam that scatters into a given solid angle $d\Omega$; as a function of direction Θ it is

$$\frac{d\sigma}{d\Omega} = \frac{2\pi b\, db}{2\pi \sin \Theta\, |d\Theta|} = \frac{b}{\sin \Theta} \left| \frac{db}{d\Theta} \right|$$

$$= \frac{1}{2} \left| \frac{d\cos \Theta}{db^2} \right|^{-1} . \tag{4.67}$$

The form of this result is the same for attractive forces, so the calculation of cross sections involves the derivative of $\cos \Theta$ with respect to b^2.

Clearly, $b(\Theta)$ is different for different force laws, since it depends explicitly on the exact motion of a particle through the area where the force is strong.

One object of scattering experiments is to deduce information about the force law from the scattering cross section. To illustrate the technique, we calculate the very important case for the inverse square, or Coulomb, force law. The result for the scattering cross section is called the **Rutherford cross section**.

In the repulsive ($k > 0$) case, we have (4.52):

$$u = u_0 \cos(\phi - \phi_0) - \frac{km}{\ell^2}, \qquad (4.68)$$

where here m is the mass of the bullet particle, since the target particle is assumed to be infinitely massive. We retain ϕ to denote the angular coordinate of the orbit; this ϕ is not the angle around the z-axis—rather, it refers to the angle with respect to the z-axis. Unlike the analysis of the preceding section, we do not take the angle of closest approach ϕ_0 to be zero; instead, the angular coordinate is determined by the initial condition that the bullet approach from $\phi = \pi$, parallel to the $(-z)$-axis.

The angle of deflection Θ is the angle ϕ when u returns to zero. It is clear from the geometry of the scattering situation that this scattering angle Θ and the angle of closest approach ϕ_0 are related by

$$\Theta = 2\phi_0 - \pi. \qquad (4.69)$$

Hence the angle ϕ_0 must be calculated as a function of the impact parameter b. Note that $\Theta = \pi - \Delta\phi$, where $\Delta\phi = \phi(t = -\infty) - \phi(t = +\infty)$.

Recall that all of the bullet particles have the same initial kinetic energy (or equivalently, the same initial speed v_0). This value of initial kinetic energy is the same as the total energy E, which appears in (4.34). When $r = r_c$, the distance of closest approach, $\dot{r} = 0$, and this equation reads

$$E = \frac{1}{2}mv_0^2 = \frac{k}{r_c} + \frac{\ell^2}{2mr_c^2}. \qquad (4.70)$$

The orbit equation (4.68) shows that at closest approach

$$\frac{1}{r_c} = u_0 - \frac{mk}{\ell^2}. \qquad (4.71)$$

We also need the expression for the angular momentum ℓ in terms of the impact parameter b:

$$\ell = mv_0 b. \qquad (4.72)$$

These relations first allow us to determine r_c and then u_0 in terms of the initial conditions, namely, the initial energy E and the impact parameter b. The result is

$$u_0 = \frac{1}{2Eb^2}\left(k^2 + 4E^2b^2\right)^{1/2}. \qquad (4.73)$$

We now return to the orbit equation (4.68), evaluated at $\phi = \pi$, where $u = 0$:

$$0 = u_0 \cos(\pi - \phi_0) - \frac{mk}{\ell^2}. \qquad (4.74)$$

Now solve this equation for $\cos(\pi - \phi_0) = (k^2 + 4E^2 b^2)^{1/2}$ using (4.69):

$$\cos(\pi - \phi_0) = \sin(\tfrac{1}{2}\Theta). \qquad (4.75)$$

We saw in (4.67) that the cross section conveniently involves $\cos\Theta$, which of course is $1 - 2\sin^2(\tfrac{1}{2}\Theta)$. Finally, the parameters in (4.74) are expressed in terms of the initial conditions, and the result is

$$\cos\Theta = \frac{4E^2 b^2 - k^2}{4E^2 b^2 + k^2}. \qquad (4.76)$$

It is now simple to calculate $d\cos\Theta/db^2$ and express it in terms of the observable Θ rather than the unobservable b. The result for the scattering cross section is

$$\frac{d\sigma}{d\Omega} = \left(\frac{k}{4E}\right)^2 (\sin\tfrac{1}{2}\Theta)^{-4}. \qquad (4.77)$$

Notice that the sign of the coefficient k in the power law does not appear in the result, so that the attractive and repulsive Coulomb interactions have the same cross section. In an experiment, the number of particles scattered by the angle Θ is measured in order to determine the cross section; thus this information cannot determine the sign of k.

One remarkable feature of the result is that it is identical to the quantum mechanical result for low-energy (i.e., nonrelativistic) Coulomb scattering of spin-zero particles. Also note that at $\Theta = \pi$ the cross section is nonzero, so that there is a finite amount of scattering to the rear. This feature led to the discovery of the atomic nucleus.

The **total cross section** σ_T is the integral of the cross section derived above (the differential cross section) over all solid angles. In the present case, since the differential cross section only depends on Θ, we have

$$\sigma_T = \int \frac{d\sigma}{d\Omega} d\Omega = 2\pi \int_0^\pi \frac{d\sigma}{d\Omega} \sin\Theta \, d\Theta. \qquad (4.78)$$

The divergent result at $\Theta = 0$ of (4.77) actually results in a divergent σ_T: Near $\Theta = 0$, this integral behaves like

$$\sigma_T \sim \int_0 \Theta^{-4} \Theta \, d\Theta \to \infty. \qquad (4.79)$$

This divergence arises from the fact that there are very many particles that pass far from the scatterer but still undergo some deflection. A force such as the Coulomb force, which has infinite total cross section due to its small-angle behavior, is said to be **long range**. If the force is nonzero at infinity, the force is long range. (In quantum mechanics, the situation is somewhat different; cf. Exercise 20.)

In principle the foregoing analysis can be carried out for all types of forces. One can start with (4.36) in order to find u as a function of ϕ or, better, ϕ as a function of u:

$$\pi - \phi_0 = \int_0^{u_c} \left[\frac{(E - V)2\mu}{\ell^2 - u^2}\right]^{-1/2} du, \qquad (4.80)$$

where the limits of the integral are the initial value of u (zero) and the value of u at closest approach u_c. The latter value is also obtained from (4.36) by setting $du/d\phi = 0$ and solving the implicit equation

$$u_c^2 + \frac{2\mu}{\ell^2}\left[V(u_c) - E\right] = 0. \tag{4.81}$$

Unfortunately, it is often the case that the integral (4.80) and the implicit Eq. (4.81) cannot be evaluated in closed form using tabulated functions.

Our analysis so far has been based on the assumption that the target particle is infinitely massive, but it is straightforward to carry out the calculation when the target can recoil. As we mentioned at the start of this section, the energy E of the bullet particle is typically so great that the target may be considered to recoil freely. The target is initially at rest; this frame is called the laboratory or **lab frame**. The analysis proceeds first, however, in the **CM frame**, in which the center of mass is fixed. The mass used in the calculations is the reduced mass μ, and the angle of scattering is Θ_{CM}, as measured from the incident direction of the bombarding particles in the CM frame. In this frame the target particle moves off or recoils at angle $\Theta_{CM} - \pi$. Since the total momentum vanishes in the CM frame, each particle has a velocity inversely proportional to its mass. In the CM frame, too, each particle has the same energy after the collision as before it.

The result in the CM frame must then be transformed into a result in the lab frame. To do so, we must determine the relation between these two frames. At the start, the target particle, of mass $m_2 < \infty$, is at rest at the origin in the lab frame. The bullet or bombarding particle, mass m_1, enters from the $(-z)$-direction with incoming energy $E = \frac{1}{2}m_1 v_0^2$ in the lab frame. The relative speed v_{CM} of the CM and lab frames is given by

$$v_{CM} = \frac{1}{M}(2m_1 E)^{1/2}, \qquad \text{with} \quad M = m_1 + m_2,$$
$$= \frac{m_1}{M}v_0. \tag{4.82}$$

In the CM frame, the bombarding particle has initial energy E_{CM} given by

$$E_{CM} = \frac{1}{2}m_1(v_0 - v_{CM})^2 = E\left(\frac{m_2}{M}\right)^2. \tag{4.83}$$

This particle scatters into the angle Θ_{CM} in the CM frame but retains its speed (since the collision is elastic). Let us assume for convenience that the bullet is initially in the $(x$–$z)$-plane, in which it stays after scattering. The x- and z-components of its final velocity after collision, therefore, are

$$v_{CM,x}^{final} = \left(\frac{2E_{CM}}{m_1}\right)^{1/2}\sin\Theta_{CM},$$
$$v_{CM,z}^{final} = \left(\frac{2E_{CM}}{m_1}\right)^{1/2}\cos\Theta_{CM}. \tag{4.84}$$

To determine the final velocity in the lab frame, v_{CM} must be added to the z-component of this velocity; we also change the parameter E_{CM} to E:

$$v_{\text{lab},x}^{\text{final}} = \left(\frac{2E}{m_1}\right)^{1/2} \frac{m_2}{M} \sin \Theta_{CM},$$

$$v_{\text{lab},z}^{\text{final}} = \left(\frac{2E}{m_1}\right)^{1/2} \frac{m_2}{M} \cos \Theta_{CM} + \left(\frac{2E}{m_1}\right)^{1/2} \frac{m_1}{M}. \tag{4.85}$$

The scattering angle in the lab frame is then

$$\Theta_{\text{lab}} = \tan^{-1} \frac{m_2 \sin \Theta_{CM}}{m_1 + m_2 \cos \Theta_{CM}}. \tag{4.86}$$

Every particle that emerges between Θ_{CM} and $\Theta_{CM} + d\Theta_{CM}$, namely, into the solid angle $d\Omega_{CM} = 2\pi \sin \Theta_{CM} |d\Theta_{CM}|$, goes through the corresponding solid angle $d\Omega_{\text{lab}} = 2\pi \sin \Theta_{\text{lab}} |d\Theta_{\text{lab}}|$. These solid angles are related through the formula (4.86) for Θ's. Hence, the differential cross section in the lab frame is

$$\frac{d\sigma}{d\Omega_{\text{lab}}} = \frac{d\sigma}{d\Omega_{CM}} \frac{d\Omega_{CM}}{d\Omega_{\text{lab}}}$$

$$= \frac{d\sigma}{d\Omega_{CM}} \frac{\sin \Theta_{CM}}{\sin \Theta_{\text{lab}}} \frac{d\Theta_{CM}}{d\Theta_{\text{lab}}}. \tag{4.87}$$

This formula gives the transformation between the center-of-mass cross section and the laboratory cross section. In general, the ratio of cross sections is somewhat complicated, but special simplifications do occur when the masses are equal.

The target particle recoils and can often be detected in the lab frame. The techniques described above may be used to calculate the laboratory cross section for the recoiling target particle. We will not give the details, however.

4.6   The Three–Body Problem

The essential reduction in the preceding sections occurs by working in the center of the mass frame. This removes three coordinates from consideration in the problem. This simplification works, of course, also in the case of three interacting bodies. However, in the three-body case the reduction is from nine coordinates (three for each body) to six ($= 9 - 3$), the three removed coordinates being those of the center of the mass. The remaining six-dimensional system is substantially complicated, and in fact analytical solutions, except in very special situations, have been impossible. An example of a known exact solution is that of three equal-mass objects orbiting in an equilateral triangular configuration under the attraction of their mutual gravitational field (Figure 4.1). It is not hard to show that the required rate of rotation is $\omega^2 = 3Gm/\ell^3$.

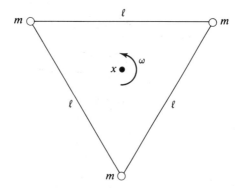

Figure 4.1

A simplification that is accurate enough as a starting point for solar system orbital dynamics is to consider the lightest of the three bodies to be a test body that has absolutely no effect on the motion of the other two bodies in the system. For instance, we may consider the three-body system to be Earth, Moon, and a spacecraft in the Earth-Moon vicinity. Then the spacecraft has certainly negligible effect on the motion of the Earth and the Moon. Other bodies in the solar system can be expected to cause perturbations of the satellite orbit, but its principal motion is determined by the already given (time-dependent) Earth-Moon gravity field. As a second example, one may consider the motion of the Earth in the field of the Sun and Jupiter. Here the mass of the Earth is not really negligible in relation to Jupiter's mass, but the orbits so determined are a good approximation and a good starting point for an approximation method.

The scheme will thus be the following. We solve completely for the motion of the two massive bodies in the system in the center-of-mass frame. This is essentially what we have already done; for gravitational attraction we can just carry over the Kepler orbits we found in the preceding section. These being known, we consider them a given time-dependent background for the motion of our test particle.

Let us try to find stationary points for the test particle of the field of the other two. To simplify further, we will assume that the large masses undergo circular motion with thus constant angular velocity ω (their motion lies in a plane). We will consider transforming to a frame centered on the center of mass but rotating at rate ω.

The Lagrangian for the test mass in the field of the other two is

$$\mathcal{L} = \tfrac{1}{2}m(\dot{r}^2 + r^2\dot{\theta}^2) - U(r,\theta,t), \tag{4.88}$$

where $u(r,\theta,t)$ is the time-dependent potential energy due to the other two masses.

If we let $\tilde{\theta} = \theta - \omega t$, we are transforming to a rotating frame. We will write the Lagrangian in rotating cylindrical coordinates $\theta = \tilde{\theta} - \omega_0 t$.

$$\mathcal{L} = \tfrac{1}{2}m\big(\dot{\rho}^2 + \rho^2(\dot{\theta} - \omega_0)^2 + \dot{z}^2\big) - \tilde{U}(x^i). \qquad (4.89)$$

(The time transformation involved is such that the time dependence has been removed from the potential.)

The complication is that two new terms appear in the Lagrangian: $-m\rho^2\dot{\theta}\omega_0$, which gives rise to a Coriolis term, and $(m/2)\rho^2\omega_0{}^2$, which is the potential for the centrifugal force. We combine this term with $\tilde{U}$ to obtain

$$\mathcal{L} = \tfrac{1}{2}m(\dot{\rho}^2 + \rho^2\dot{\theta}^2 + \dot{z}^2) - \omega_0 m\rho^2\dot{\theta} - U_{\rm rot}(x^i). \qquad (4.90)$$

This form has several advantages, particularly for the consideration of stable co-rotating orbits. These are orbits with $\dot{\rho} = \dot{\theta} = \dot{z} = 0$. These are also *minima* of the function $U_{\rm rot}(x^i)$. We shall find some such orbits, as well as some unstable ones.

First, note that there are no such orbits with $z \neq 0$. A point with $\dot{x}^i = 0$ but $z > 0$ will always be attracted downward; with $z < 0$, it will always be attracted upward. Hence $z = 0$ test orbits will oscillate through the plane defined by the orbit(s) of the two massive bodies. Co-rotating orbits must therefore lie in the plane. We can find some of these orbits by symmetry.

For instance, consider the three masses aligned along the Earth-Moon line. There is in principle an equilibrium at the point on this line where the net gravitation force and the centrifugal force cancel one another. There are in fact three such positions. One of them lies between the two massive bodies. The force per unit mass on an object located between them is

$$\frac{F}{m} = \frac{GM_\oplus}{r_1{}^2} - \frac{GM_{\mathbb{C}}}{r_2{}^2} - r_3\omega^2 \qquad (4.91)$$

(toward the Earth). Here r_1 is the distance from the Earth to the test mass, r_2 is the distance from the Moon to the test mass, and r_3 is the distance from the Earth-Moon center of mass to the test object. As can be seen from Figure 4.2, at some point nearer the moon than the CM, we expect centrifugal force plus lunar gravity to exceed the attraction of Earth gravitation. Define $r = r_\oplus + r_{\mathbb{C}}$, which is the total Earth-Moon separation distance. Also, $r_1 = r_\oplus + r_3$ and $r_2 = r_{\mathbb{C}} - r_3$. The quantity ω^2 is given by

$$\omega^2 = \frac{G(M_\oplus + M_{\mathbb{C}})}{r^3}.$$

The location of the mass m can be found by equating the force in Eq. (4.91) to zero:

$$0 = M_\oplus(r_{\mathbb{C}} - r_3)^2 - M_{\mathbb{C}}(r_\oplus + r_3)^2 - \omega^2(r_{\mathbb{C}} - r_3)^2(r_\oplus - r_3)^2 r_3. \qquad (4.92)$$

This is a genuine quintic equation, which can be solved only numerically for the location r_3 of the equilibrium points. There are three real solutions and one complex pair. The three real solutions correspond to positions that can be predicted by considering the attractive gravity/centrifugal force balance. Besides the solution sketched in Figure 4.2, there are two points lying on the line but

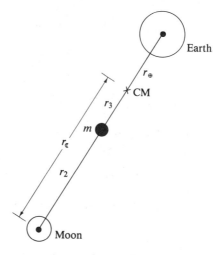

Figure 4.2

outside the two masses. Consider that each large mass is in a circular orbit because the force per unit mass on it from the other large mass just balances the centrifugal force on it. Outside its orbital position, the centrifugal force will be greater, but a test body placed there will also feel a stronger inward gravitational force (because it is reacting to the sum of *two* inward gravitational sources).

The labeling of these points is as shown in Figure 4.3. Each of the points L_1, L_2, and L_3 is an unstable equilibrium point. Consider: Moving L_3 farther out will reduce the inward gravitational attraction and increase the ωr^2 centrifugal term so that it will accelerate outward; similarly for L_2. Moving L_1 toward the Earth will increase the attractive force toward the Earth and reduce

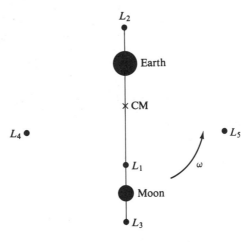

Figure 4.3

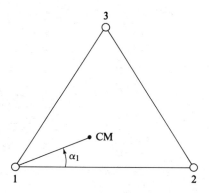

Figure 4.4

the centrifugal and Moon-attractive forces so that it will accelerate toward the Earth.

To proceed, we reconsider the equilateral triangle solution introduced at the beginning of this section. Here we take all three masses to be significant (not test masses) and in principle different from one another. Label the mass points ①, ②, and ③; we will show that there exists an equilateral triangle orbit, even in the unequal non-test-mass case.

Consider an instant when the masses are aligned with M_1 at the origin, and with the M_1–M_2 axis aligned along the x-axis. Figure 4.4 shows this situation. The location of the center of mass is then

$$X_{\text{CM}} = \ell\,\frac{M_2 + M_3/2}{M} \qquad Y_{\text{CM}} = \frac{\ell M_3 \sqrt{3}}{2M}$$

where ℓ is the length of each side of the triangle; and $r_1 = (X_{\text{CM}}^2 + Y_{\text{CM}}^2)^{1/2}$ is the distance from M_1 to the center of mass.

If we are to have a steady-state rotating solution, the net force on M_1 must be through the center of mass, and the force equation on M_1 must read

$$0 = \frac{F_1}{M_1} = r_1\omega^2 - \frac{GM_2}{\ell^2}\cos\alpha_1 + \frac{GM_3}{\ell^2}\cos(60° - \alpha_1). \qquad (4.93)$$

Using the formula for the cosine of the sum of angles, along with $\cos\alpha_1 = X_{\text{CM}}/r_1$; $\sin\alpha_1 = Y_{\text{CM}}/r_1$, we obtain

$$\omega^2 \ell\,\frac{(M_2^2 + M_3^2 + M_2 M_3)^{1/2}}{M} =$$

$$= \frac{G}{\ell^2}\left[M_2\left(M_2 + \frac{M_3}{2}\right) + \frac{M_3}{2}\left(M_2 + \frac{M_3}{2}\right) + \left(\frac{M_3\sqrt{3}}{2}\right)^2 \right] \times \qquad (4.94)$$

$$\times (M_2^2 + M_3^2 + M_2 M_3)^{-1/2},$$

which simplifies to

$$\omega^2 = \frac{GM}{\ell^3}. \tag{4.95}$$

With this choice for ω, we also find that the forces transverse to the line through the center of mass sum to zero; M_1 will remain on its circular orbit. Because nothing is special about M_1, we could have performed this equation based on any one of the masses, so we conclude that it is satisfied for each and we have a general solution for this equilateral triangle kind of circular motion. Because we have this solution in general, it certainly holds if any one of the three masses is an infinitesimal test mass. We thus have the two additional Lagrange points L_4, L_5 (see Figure 4.3). These are also called Trojan points because the Trojan asteroid group orbits in L_4, L_5 in the Sun-Jupiter system.

The statement that the Trojan asteroids exist gives observational evidence to the stability of test-body orbits near L_4 and L_5. In fact, the stability of L_4, L_5 orbits depends on the ratio of the masses in the system. Remarkably, it is possible to do an analytical treatment of the frequencies of motion near the equilibrium points of the masses, even if all the masses are significant. The behavior is $\delta x \sim e^{\lambda t}$, where imaginary λ means oscillatory motion near the equilibrium points. It is found that there are frequencies associated with the motion of the form

$$\lambda_{\pm}^2 = -\frac{\omega^2}{2}\left[1 \pm \left(\frac{1 - 27(M_1 M_2 + M_2 M_3 + M_3 M_1)}{M^2}\right)^{1/2}\right]. \tag{4.96}$$

If $\lambda_{\pm}^2$ is real and negative, we have oscillatory motion, but if the square root yields an imaginary number, two of the four roots for λ have exponentially growing solutions; in other words, stability requires that

$$(M_1 + M_2 + M_3)^2 > 27(M_1 M_2 + M_2 M_3 + M_1 M_3). \tag{4.97}$$

Suppose that

$$M_3 > M_2 > M_1 \sim 0;$$

that is, M_3 is the largest and M_1 is a test mass. Then the condition for stability is

$$(M_2 + M_3)^2 > 27 M_2 M_3$$

or

$$M_3 \gtrsim 24.96 M_2. \tag{4.98}$$

This condition is satisfied for the Sun-Jupiter system so that the Trojan orbits are stable. It is also satisfied for $M_3 = M_{\oplus}, M_2 = M_{\mathbb{C}}$, so that L_4 and L_5 are stable test-body orbits in the Earth-Moon system. Perturbations from other solar system bodies may affect the long-term stability of these orbits, however.

EXERCISES

4.1. The harmonic oscillator in two dimensions is called degenerate if its frequencies of oscillation in the two coordinate x, y directions are the same. Its kinetic energy is

$$T = \tfrac{1}{2}m(v_x{}^2 + v_y{}^2),$$

and its potential energy is

$$V = \tfrac{1}{2}k(x^2 + y^2),$$

where m is the mass, k is a spring constant, and v_x and v_y are velocity components.

(a) Construct the Lagrangian L and verify that the Euler-Lagrange equations are equivalent to Newton's second law.

(b) A second Lagrangian for this system is $L' = v_x v_y - (k/m)xy$. Prove this assertion by showing that the Euler-Lagrange equations derived from L', although not identical to those derived in part (a), are so similar that they obviously have the same set of solutions.

(c) Now transform the coordinates $\{x, y\}$ to the plane-cylindrical coordinates $\{r, \phi\}$ in L. Again obtain the equations of motion. Prove that in this coordinate system, one of the coordinates is cyclic.

4.2. (a) Prove (4.6) that

$$\sum_{a=1}^{N} m_a \, \xi^i{}_a = 0.$$

(b) Then show (4.7) that

$$T = \tfrac{1}{2}M\delta_{ij}U^iU^j + \tfrac{1}{2}\sum_{a=1}^{N} m_a \, \delta_{ij} \, w^i{}_a w^j{}_a.$$

4.3. (a) Verify Eqs. (4.15)–(4.16) for the two-body problem.

(b) In the central force problem for two particles, the three components of the angular momentum vector ℓ_a are

$$\ell_1 = \mu(q^2 v^3 - q^3 v^2),$$
$$\ell_2 = \mu(q^3 v^1 - q^1 v^3),$$
$$\ell_3 = \mu(q^1 v^2 - q^2 v^1).$$

Express these three components as functions of r, θ, ϕ after verifying the general expression above. Note that this is not the same as finding the components of the angular momentum vector after a change of coordinates to $\{r, \theta, \phi\}$; it is simply the expression of the components in the $\{q^k\}$ coordinate system as three functions of other variables. Show that the condition $\ell_1 = \ell_2 = 0$ is equivalent to $\theta = \pi/2 =$ const . (under the general presumption that r, θ, ϕ do not vanish).

4.4. Use any good reference source (e.g., C. W. Allen, *Astrophysical Quantities*, 3rd edition, Athlone Press, London, 1973) to obtain the size and mass of the sun and the mass and (typical) orbital radii of the planets. Describe the location of the center of mass of the solar system. Is it *ever* at the center of the sun? What is the maximum distance between the center of the sun and center of mass of the solar system?

4.5. If we consider motion under an attractive central force, circular orbits may be possible.
   (a) What can you say about the types of potential functions that allow circular orbits? That is, under what conditions are circular orbits possible?
   (b) For circular orbits in an attractive central force potential of the form $V = -ar^{-n}$, find a relation between the kinetic and the potential energies. Show that the same relation holds between the average values of the kinetic energy and the potential energy even in the case of noncircular orbits.

4.6. Derive **Kepler's third law**, the relation between the period of a closed orbit in an attractive inverse square central force and the size (semimajor axis, which is half the largest diameter) of the orbit. The relation involves the strength of the central force.

4.7. A spaceship is in a circular orbit with a 4-hour period around the Earth. A mutiny occurs, and the captain (in a spacesuit) is bodily thrown out of the side hatch (straight toward the center of the Earth). Will the captain ever meet up with his ship again? If so, when?

4.8. A star is seen orbiting an invisible companion. The mass of the visible star is known, as is the distance to the star. The only things that can be measured concerning the orbit are the observed positions of the visible star at various times; the inclination of the orbit to the line of sight cannot be determined. (For example, a circular orbit might appear elliptical.) Can the mass of the invisible companion be determined? If so, show how.

4.9. Suppose that the Earth were exactly spherical and of constant density ρ inside. Poisson's law determines the potential function $V(r)$ by

$$\nabla^2 V = 4\pi G \rho,$$

where G is Newton's constant of gravitation. Note that in spherical coordinates $\{r, \theta, \phi\}$, the Laplace operator acts on a function that is only a function of r in the following way:

$$\nabla^2 V = \frac{1}{r} \frac{d^2(rV)}{dr^2}.$$

Obtain the potential $V(r)$ at all points inside and outside the surface of the Earth. Using this potential, calculate the orbits of particles that can pass through the Earth unaffected except for their gravitational interaction (a massive neutrino moving nonrelativistically, if it exists, for example).

What is the relation between circular orbits just below and just above the surface? Describe the noncircular orbits of particles that stay below the surface.

4.10. The orbit of Mercury is an approximate ellipse whose perihelion is observed to advance by the small precession amount of 42 minutes of arc per century. One might hope to model this effect by a small inverse cube force addition to the inverse square force law that would result from a spherical sun. By straightforward calculation, show from the Poisson equation of Exercise 4.9 that if distributed in a spherically symmetric manner, an unacceptably large amount of mass must be present in a matter distribution that gives rise to the inverse cube force necessary to explain the precession of Mercury.

4.11. In the $\{q^i\}$ coordinates of the two-body problem, define the vector C_i by

$$C_i = \varepsilon_{ijk}\, v^j \ell^k - \frac{k}{r}\, q^j \delta_{ij},$$

where ℓ^k are the components of the angular momentum. Suppose that the force is an inverse square force.
(a) Show that C_i is conserved.
(b) Show that $C_i\, \ell^i = 0$; deduce that C_i must lie in the plane of the orbit.
(c) C_i, ℓ^i, and the energy E are constants of the motion for inverse square forces. Show that of the seven components listed, only five are independent.
(d) Rewrite the definition of C_i as

$$\frac{k}{r}\, q^j \delta_{ij} = \varepsilon_{ijk}\, v^j \ell^k - C_i,$$

and square both sides. The resulting equation involves only v^j and not q^i, since the left side is k times a unit vector. Use the fact that the motion is in a plane to show that this equation is the equation of a circle. Conclude that although the orbit is elliptical in space, the motion is a circle in the space of velocities.

4.12. Give the generic solution (involving two constants of integration) for the nonlinear oscillator equation

$$m\ddot{x} + kx - \frac{A}{x^3} = 0$$

for $A > 0$. Now describe a physical situation in which a similar equation is needed to find the orbit of one particle around another.

4.13. Consider the elastic scattering of point particles from an infinitely massive, perfectly hard sphere (where the angle of incidence equals the angle of reflection at the point of impact). Calculate the differential cross section and verify that the total cross section is the same as the geometric cross section.

4.14. A scattering experiment can be carried out in an inverse square force field by aiming asteroids at the moon. Because the moon has a nonzero size, there is a finite probability that an asteroid will collide with the moon. Assume that any such asteroid will be absorbed onto the surface and not bounce off. Use the actual mass and radius of the moon (but assume that it is a perfect sphere) and determine the absorption cross section as a function of the initial velocity of the asteroids. (Why is it not a function of the mass of the asteroids?)

4.15. Newtonian calculations can often give valuable intuitive insight into relativistic situations and can give simple order-of-magnitude estimates of quantities that may be complicated to calculate in full relativistic detail. For example, one may calculate the deflection of light by the gravitational field of the sun by assuming that light consists of corpuscles moving at speed c. Einstein did this later calculation and showed that the result was off by only a factor of 2. Calculate the following quantities: Find the function $\theta(b)$ or $b(\theta)$, and obtain the differential cross section for scattering of corpuscles moving at speed c past the sun. Insert the actual values of the radius and mass of the sun in order to find the maximum deflection for a ray that just grazes the sun. Find the absorption cross section as in Exercise 4.14.

4.16. The law of refraction in optics is Snell's law:

$$n_1 \sin \theta_1 = n_2 \sin \theta_2,$$

where θ_1 and θ_2 are the angles of incidence and refraction (measured with respect to the perpendicular to the boundary), and n_1 and n_2 are the indices of refraction on the two sides of the boundary. Calculate the differential and total cross sections for scattering of light by a glass sphere with index of refraction 1.5 in a vacuum. (Neglect all reflections.)

4.17. A perfectly reflecting ellipsoid of revolution is put in a beam of light. The beam is incident along the axis of symmetry of the ellipsoid. What is the scattering cross section for the system? What is the total cross section?

4.18. A scattering cross section does not uniquely determine the force law that produces it. For instance, attractive and repulsive Coulomb scattering produce the same results. Furthermore, for instance, the Rutherford cross section for scattering by an inverse square force law may be equal to the scattering cross section due to a different force law entirely, even a nonspherical force law. For this exercise, consider the scattering due to a single, infinitely massive center for an inverse square force compared to elastic scattering off a surface.

(a) Consider a surface of revolution that is oriented so that incoming particles strike it from a direction parallel to its axis of symmetry. Suppose that these particles bounce off elastically from its surface. Find a surface (which will be infinite in extent) such that the cross section for elastic scattering is the same as the Rutherford cross section.

(b) Using the results of part (a) as motivation, show that there is a lacuna or blank volume in the scattering problem for the repulsive inverse square force law (i.e., a volume where particles do not penetrate). Verify the existence of this shadow region by considering the actual orbits of particles in the repulsive inverse square force field and the caustic surface formed by them. Show how the effect of the shadow appears in the form of the Rutherford cross section.

4.19. Obtain the differential and total scattering cross section for particles scattering in a fixed, repulsive inverse cube central force $F \propto r^{-3}$.

4.20. The Yukawa potential has the form

$$V = \frac{1}{r} k e^{-\lambda r}.$$

Show how to find the differential and total cross sections for this potential. Is the force derived from this potential long range or not? [*Answer:* long range; in quantum mechanics this potential is short range (i.e., produces a finite total cross section).]

4.21. Consider the potential

$$V = k r^{-\alpha},$$

where α is a positive constant, not necessarily an integer. By considering the total cross section, show that for all values of α the force derived from V is long range (produces an infinite cross section).

5

EQUIVALENT LAGRANGIANS

Let $L(q^i, \dot{q}^i, t)$ be a Lagrangian. A second Lagrangian may be formed from L by multiplying L by a nonzero constant and adding a total time derivative of an arbitrary function of q^i, t (not of $\dot{q}^i$). The second Lagrangian yields the same equations of motion as the first; one computes $\dot{f} = (\partial f / \partial q^i)\dot{q}^i + \partial f / \partial t$ and finds that

$$\frac{d(\partial \dot{f}/\partial \dot{q}^i)}{dt} - \frac{\partial \dot{f}}{\partial q^i} = 0. \tag{5.1}$$

Now let $L(q^i, \dot{q}^i, t)$ be a Lagrangian, and let k and $f(q^i, t)$ be an arbitrary nonzero constant and function, respectively. Define $\bar{L}$ by

$$\bar{L}(q^i, \dot{q}^i, t) = kL + \dot{f}. \tag{5.2}$$

Show that $\bar{L}$ and L give equivalent equations of motion (more precisely, the Euler-Lagrange equations formed from $\bar{L}$ are proportional to those from L; therefore, the two sets of equations are exactly equivalent). (Exercise 5.1 is to prove these results.)

More generally, we ask when we can find a second Lagrangian $\bar{L}(q^i, \dot{q}^i, t)$ that yields equations of motion having exactly the same set of solutions as those of the equations of motion formed from the first Lagrangian L. We define

$$L_i \equiv \left(\frac{\partial L}{\partial \dot{q}^i}\right)^{\cdot} - \frac{\partial L}{\partial q^i},$$

$$\bar{L}_i \equiv \left(\frac{\partial \bar{L}}{\partial \dot{q}^i}\right)^{\cdot} - \frac{\partial \bar{L}}{\partial q^i}. \tag{5.3}$$

The equations $L_i = 0$ are equivalent to $\bar{L}_i = 0$ when

$$\bar{L}_i = \Lambda_i^j L_j, \tag{5.4}$$

where the matrix Λ_i^j is a nonsingular function of $q^i, \dot{q}^i, t$:

$$\det \Lambda \neq 0. \tag{5.5}$$

We then say that L and $\bar{L}$ are **equivalent**. The notation **s-equivalent** is also used, the s meaning that the solutions to the equations $L_i = 0$ and $\bar{L}_i = 0$ are the same. Of course, it may happen that $\det \Lambda = 0$ for particular values of $q^i, \dot{q}^i, t$. The equation $\det \Lambda = 0$ is a differential equation, and if the solutions are but special cases of $L_i = 0$, then L and $\bar{L}$ can still be called equivalent.

In this chapter we describe some of the theory of equivalent Lagrangians, a theory that allows Lagrangians not at all related by the simple multiplication by a constant or the addition of a total time derivative. One of the more interesting consequences of the nonuniqueness of Lagrangians is the effect on quantum theory. It is often the case that two different equivalent Lagrangians give quite different quantum theories. Detailed discussion of this subject is beyond the scope of this book, although we will give an example at the end of this section.

5.1 The Free Particle

Before delving into the general theory of inequivalent Lagrangians, let us examine in detail the free particle in three dimensions as an example. The coordinates of the particle are q^i $(i = 1, 2, 3)$, and we take the mass to be unity. The equations of motion are given by Newton's laws:

$$\ddot{q}^i = 0. \tag{5.6}$$

The conventional Lagrangian that yields these equations is

$$L = \tfrac{1}{2} \delta_{ij} \dot{q}^i \dot{q}^j. \tag{5.7}$$

If δ_{ij} is replaced by any constant, nonsingular matrix, the new Lagrangian will yield the same equations of motion (or equivalent ones). The new matrix need not even be positive definite.

An equivalent Lagrangian $\bar{L}$ may be, first of all, defined as a general function of $\dot{q}^i$: $\bar{L} = \bar{L}(\dot{q}^i)$. The equations of motion are

$$\bar{L}_i = \bar{W}_{ij}\,\ddot{q}^i = 0, \qquad \text{where} \quad \bar{W}_{ij} = \frac{\partial^2 \bar{L}}{\partial \dot{q}^i\,\partial \dot{q}^j}. \tag{5.8}$$

Of course, we require that $\bar{W}_{ij}$ be a nonsingular matrix except possibly at discrete values of the $\dot{q}^i$. These values are where the $\dot{q}^i$ are equal to particular values for the solutions of the equations $\ddot{q}^i = 0$. (Exercise 5.2 describes a special case.)

In the general case, $\bar{W}_{ij}$ is a function only of $\dot{q}^i$, which is a constant of the motion. Therefore, the matrix Λ_i^j (which here equals $\bar{W}_{ij}$) is composed of constants of the motion. Therefore, the trace of any power of Λ_i^j is a constant of the motion. The latter property, that any scalar formed from Λ_i^j is a constant of the motion, holds whenever we have a Λ_i^j defined by equivalent Lagrangians.

Constants of the motion play important roles in this theory. For the free particle in one dimension, with coordinate $q(t)$, we have that $c_1 = \dot{q}$ and $c_2 = q - \dot{q}t$ are constants of the motion. (Thus c_2 is a time-dependent constant!) Moreover, any constant of the motion is a function of c_1 and c_2 (see Exercise 5.3).

In continuing our study of the free particle as an example, we concentrate our attention on the free particle in one dimension.

Theorem

Let $\bar{L}(q, \dot{q}, t)$ be a Lagrangian that gives an equation of motion equivalent to the free particle (since $i = 1$ only, $\bar{L}_1$ is our notation for the equation of motion):

$$\bar{L}_1 = \left(\frac{\partial \bar{L}}{\partial \dot{q}}\right)^{\!\cdot} - \frac{\partial \bar{L}}{\partial q} = \Lambda \ddot{q}. \tag{5.9}$$

Then $\Lambda = \Lambda(q, \dot{q}, t)$ is a constant of the motion.

The proof of this fact starts with the equation of motion:

$$\bar{L}_1 = \frac{\partial^2 \bar{L}}{\partial \dot{q}^2}\,\ddot{q} + \frac{\partial^2 L}{\partial \dot{q}\,\partial q}\,\dot{q} + \frac{\partial^2 \bar{L}}{\partial \dot{q}\,\partial t} - \frac{\partial \bar{L}}{\partial q}, \tag{5.10}$$

so that $\tilde{L}_1 = \Lambda \ddot{q}$ requires

$$\Lambda = \frac{\partial^2 \bar{L}}{\partial \dot{q}^2} \tag{5.11}$$

and

$$\frac{\partial^2 \bar{L}}{\partial \dot{q}\,\partial q}\,\dot{q} + \frac{\partial^2 \bar{L}}{\partial \dot{q}\,\partial t} - \frac{\partial \bar{L}}{\partial q} = 0. \tag{5.12}$$

Note that these equations hold identically, that is, whether the equation of motion ($\ddot{q} = 0$) is satisfied or not. We differentiate the last equation with respect to $\dot{q}$:

$$\frac{\partial^3 \bar{L}}{\partial \dot{q}^2 \, \partial q} \dot{q} + \frac{\partial^3 \bar{L}}{\partial \dot{q}^2 \, \partial t} = 0. \tag{5.13}$$

We now compute

$$\frac{d\Lambda}{dt} = \frac{\partial^3 \bar{L}}{\partial \dot{q}^3} \ddot{q} + \frac{\partial^3 \bar{L}}{\partial \dot{q}^2 \, \partial q} \dot{q} + \frac{\partial^3 \bar{L}}{\partial \dot{q}^2 \, \partial t}. \tag{5.14}$$

The last two terms thus vanish identically by (5.13), and the first vanishes if the equation of motion holds. Therefore, $d\Lambda/dt = 0$.

Thus Λ is a constant of the motion, and it therefore must be of the form

$$\Lambda = \Lambda(c_1, c_2) \tag{5.15}$$

because of Exercise 5.3. We therefore ask: Does an $\bar{L}$ exist that gives an equation of motion $\Lambda \ddot{q} = 0$ for a prechosen Λ? The answer is yes. We will do an explicit example first (and give an exercise) before treating the general case.

As an example, let $\Lambda = c_2 = q - \dot{q}t$. It is then easy to show that there is at least one $\bar{L}$ such that

$$\bar{L}_1 = \Lambda \ddot{q}. \tag{5.16}$$

To show this fact, recall that

$$\Lambda = \frac{\partial^2 \bar{L}}{\partial \dot{q}^2} = c_2 = q - \dot{q}t. \tag{5.17}$$

The most general solution of this equation is

$$\bar{L} = \tfrac{1}{2} q \dot{q}^2 - \tfrac{1}{6} t \dot{q}^3 + F(q, t)\dot{q} + G(q, t). \tag{5.18}$$

To determine F and G, we now turn to the equation [Eq. (5.10) with now $\ddot{q} = 0$ assumed]

$$\frac{\partial^2 \bar{L}}{\partial \dot{q} \, \partial q} \dot{q} + \frac{\partial^2 \bar{L}}{\partial \dot{q} \, \partial t} - \frac{\partial \bar{L}}{\partial q} = 0$$

$$= \dot{q}^2 + \frac{\partial F}{\partial q} \dot{q} - \frac{1}{2} \dot{q}^2 + \frac{\partial F}{\partial t} - \frac{1}{2} \dot{q}^2 - \frac{\partial F}{\partial q} \dot{q} - \frac{\partial G}{\partial q}$$

$$= \frac{\partial F}{\partial t} - \frac{\partial G}{\partial q}. \tag{5.19}$$

We see that one solution is $F = G = 0$. The most general solution is to give $G(q, t)$ and determine F by integrating with respect to t.

In the example above, if G is given arbitrarily and F is then determined, it is easy to show (see Exercise 5.4) that the combination $F\dot{q} + G$ is the total time derivative of some function $Z(q, t)$. Since a total time derivative is really

not a significant addition to a Lagrangian, the solution for this example is best
taken as the Lagrangian

$$\bar{L} = \tfrac{1}{2}q\dot{q}^2 - \tfrac{1}{6}t\dot{q}^3. \tag{5.20}$$

5.2   One-Dimensional Case

The case of a particle moving in one dimension can be treated thoroughly; the
multidimensional cases still have many open questions. We first consider the
situation when a single coordinate $q(t)$ is involved and will but briefly describe
the higher dimensions later. Let $L(q, \dot{q}, t)$ be the Lagrangian, and write L_1 for
the equation of motion:

$$
\begin{aligned}
L_1 &= \left(\frac{\partial L}{\partial \dot{q}}\right)^{\cdot} - \frac{\partial L}{\partial q} \\
&= \frac{\partial^2 L}{\partial \dot{q}^2}\,\ddot{q} + \frac{\partial^2 L}{\partial \dot{q}\,\partial q}\,\dot{q} + \frac{\partial^2 L}{\partial \dot{q}\,\partial t} - \frac{\partial L}{\partial q} = 0.
\end{aligned}
\tag{5.21}
$$

We presume that the coefficient of $\ddot{q}$ is nonzero, so that we may write

$$\ddot{q} = \left(\frac{\partial^2 L}{\partial \dot{q}^2}\right)^{-1}\left[-\frac{\partial^2 L}{\partial \dot{q}\,\partial q}\,\dot{q} - \frac{\partial^2 L}{\partial \dot{q}\,\partial t} + \frac{\partial L}{\partial q}\right]. \tag{5.22}$$

We seek another Lagrangian $\bar{L}$ whose equation of motion is $\bar{L}_1 = 0$. This
second equation of motion, too, will be written

$$\ddot{q} = \left(\frac{\partial^2 \bar{L}}{\partial \dot{q}^2}\right)^{-1}\left[-\frac{\partial^2 \bar{L}}{\partial \dot{q}\,\partial q}\,\dot{q} - \frac{\partial^2 \bar{L}}{\partial \dot{q}\,\partial t} + \frac{\partial \bar{L}}{\partial q}\right]. \tag{5.23}$$

The two Lagrangians L and $\bar{L}$ are equivalent if their equations of motion have
the same solutions. This equivalence can happen only when the right sides of
the two $\ddot{q}$ equations are identically equal as functions of $q, \dot{q}, t$. We define Λ by

$$\Lambda \equiv \frac{\partial^2 \bar{L}}{\partial \dot{q}^2}\left(\frac{\partial^2 L}{\partial \dot{q}^2}\right)^{-1}, \tag{5.24}$$

so that

$$
\begin{aligned}
\frac{\partial^2 \bar{L}}{\partial \dot{q}\,\partial q}\,\dot{q} &+ \left(\frac{\partial^2 \bar{L}}{\partial \dot{q}\,\partial t}\right) - \frac{\partial \bar{L}}{\partial q} \\
&= \Lambda\left(\frac{\partial^2 L}{\partial \dot{q}\,\partial q}\,\dot{q} + \frac{\partial^2 L}{\partial \dot{q}\,\partial t} - \frac{\partial L}{\partial q}\right).
\end{aligned}
\tag{5.25}
$$

This equation must hold identically whenever q is a solution of the equation
of motion because we can pick initial data that spans configuration and tangent
space. We can also see that

$$\bar{L}_1 = \Lambda L_1. \tag{5.26}$$

We first prove that Λ is a constant of the motion: First, we have

$$\frac{d\Lambda}{dt} = \frac{\partial \Lambda}{\partial \dot{q}}\,\ddot{q} + \frac{\partial \Lambda}{\partial q}\,\dot{q} + \frac{\partial \Lambda}{\partial t}. \tag{5.27}$$

We do not assume that $q(t)$ solves the equation of motion derived from L, but we do substitute

$$\ddot{q} = \left(\frac{\partial^2 L}{\partial \dot{q}^2}\right)^{-1}\left(-\frac{\partial^2 L}{\partial \dot{q}\,\partial q}\,\dot{q} - \frac{\partial^2 L}{\partial \dot{q}\,\partial t} + \frac{\partial L}{\partial q}\right). \tag{5.28}$$

We also use the definition of Λ for the last two terms in $d\Lambda/dt$:

$$\begin{aligned}
\frac{d\Lambda}{dt} &= \frac{\partial \Lambda}{\partial \dot{q}}\left(\frac{\partial^2 L}{\partial \dot{q}^2}\right)^{-1}\left(L_1 - \frac{\partial^2 L}{\partial \dot{q}\,\partial q}\,\dot{q} - \frac{\partial^2 L}{\partial \dot{q}\,\partial t} + \frac{\partial L}{\partial q}\right) \\
&\quad - \left(\frac{\partial^2 L}{\partial \dot{q}^2}\right)^{-2}\frac{\partial^2 \bar{L}}{\partial \dot{q}^2}\left(\frac{\partial^3 L}{\partial \dot{q}^2\,\partial q}\,\dot{q} + \frac{\partial^3 L}{\partial \dot{q}^2\,\partial t}\right) \\
&\quad + \left(\frac{\partial^2 L}{\partial \dot{q}^2}\right)^{-1}\left(\frac{\partial^3 \bar{L}}{\partial \dot{q}^2\,\partial q}\,\dot{q} + \frac{\partial^3 \bar{L}}{\partial \dot{q}^2\,\partial t}\right).
\end{aligned} \tag{5.29}$$

We now take the $\dot{q}$ derivative of Eq. (5.25) that results from the equality of the formulas for $\ddot{q}$ from L and $\bar{L}$:

$$\begin{aligned}
\frac{\partial^3 \bar{L}}{\partial \dot{q}^2\,\partial q}\,\dot{q} + \frac{\partial^3 \bar{L}}{\partial \dot{q}^2\,\partial t} &= \frac{\partial \Lambda}{\partial \dot{q}}\left(\frac{\partial^2 L}{\partial \dot{q}\,\partial q}\,\dot{q} + \frac{\partial^2 L}{\partial \dot{q}\,\partial t} - \frac{\partial L}{\partial q}\right) \\
&\quad + \Lambda\left(\frac{\partial^3 L}{\partial \dot{q}^2\,\partial q}\,\dot{q} + \frac{\partial^3 L}{\partial \dot{q}^2\,\partial t}\right).
\end{aligned} \tag{5.30}$$

By using this equation (be sure to check this result for yourself!) we see that

$$\frac{d\Lambda}{dt} = \left(\frac{\partial \Lambda}{\partial \dot{q}}\right)^{-1} L_1 \tag{5.31}$$

identically (whether $L_1 = 0$ or not). Now, if L_1 does vanish, so that $q(t)$ is a solution of the equation of motion, we see that Λ is a constant of the motion. (A similar result holds in the multidimensional case.)

Conversely, if we choose any constant of the motion Λ, there does exist a Lagrangian $\bar{L}$ such that $\bar{L}_1 = \Lambda L_1$. (There is at present no similar result for multidimensional systems.) The new $\bar{L}$ is unique up to the addition of a total time derivative. If $\Lambda = k = \text{const}$ (a true numerical constant), then $\bar{L} = kL$ (up to a term $\dot{f}$). The function Λ is said to "foul" the Lagrangian L or to be a **fouling function**.

We prove the remarks above, partially by reference to an exercise. First, as to the uniqueness of $\bar{L}$: If $\bar{L}$ and $\bar{\bar{L}}$ result from the same fouling function Λ, then

$$\bar{L}_1 = \bar{\bar{L}}_1 = L_1. \tag{5.32}$$

To prove the uniqueness, it is necessary to show that if $\bar{L}_1 = \bar{\bar{L}}_1$ identically, then $\bar{L} - \bar{\bar{L}}$ is a total time derivative. (This is the converse of Exercise 5.1.) For convenience, we drop the bars on the second Lagrangian and restate this result as a statement involving $\Lambda = 1$ (to be proved in Exercise 5.6): If $\Lambda = 1$, there exists a function $f(q, t)$ such that $\bar{L} - L = df/dt$.

We now show that there is at least one $\bar{L}$ such that $\bar{L}_1 = \Lambda L_1$ for any given constant of the motion Λ. To show this fact, we first solve

$$\frac{\partial^2 \bar{L}}{\partial \dot{q}^2} = \Lambda \frac{\partial^2 L}{\partial \dot{q}^2}. \tag{5.33}$$

The solution involves a double integration with respect to $d\dot{q}$, but it is easy to convert this double integration into two single ones using integration by parts.

Let $F(x)$ be any function. Then

$$\int^x dx' \int^{x'} dx'' \, F(x'') = x \int^x dx' \, F(x') - \int^x dx' \, x' \, F(x'). \tag{5.34}$$

[These are indefinite integrals. The proof is left to the reader; as a hint, work with $G(x) = \int^x dx' F(x')$.] Then the general solution for $\bar{L}$ from Eq. (5.34) is

$$\bar{L} = \dot{q} \int_c^{\dot{q}} d\dot{q}' \, \Lambda \frac{\partial^2 L}{\partial \dot{q}'^2} - \int_c^{\dot{q}} d\dot{q}' \, \dot{q}' \, \Lambda \frac{\partial^2 L}{\partial \dot{q}'^2} + A(q, t)\dot{q} + B(q, t). \tag{5.35}$$

The lower limits on the integrals may be functions of q, t, but we will here assume they are constant. [Exercise 5.7 is to carry through the calculations below assuming that $c = c(q, t)$.] We also have that $d\Lambda/dt = 0$, and this implies that

$$\frac{\partial \Lambda}{\partial t} + \frac{\partial \Lambda}{\partial q}\dot{q} = -\frac{\partial \Lambda}{\partial \dot{q}}\ddot{q} = \frac{\partial \Lambda}{\partial \dot{q}}\left(\frac{\partial^2 L}{\partial \dot{q}^2}\right)^{-1}\left(\frac{\partial^2 L}{\partial q \, \partial \dot{q}}\dot{q} + \frac{\partial^2 L}{\partial t \, \partial \dot{q}} - \frac{\partial L}{\partial q}\right). \tag{5.36}$$

To find the functions A and B, we substitute the expression (5.35) for $\bar{L}$ into the basic expression guaranteeing the equation for $\ddot{q}$, Eq. (5.25):

$$\frac{\partial^2 \bar{L}}{\partial q \, \partial \dot{q}}\dot{q} + \frac{\partial^2 \bar{L}}{\partial t \, \partial \dot{q}} - \frac{\partial \bar{L}}{\partial q}$$

$$= \Lambda\left(\frac{\partial^2 L}{\partial q \, \partial \dot{q}}\dot{q} + \frac{\partial^2 L}{\partial t \, \partial \dot{q}} - \frac{\partial L}{\partial q}\right)$$

$$= \int_c^{\dot{q}}\left(\frac{\partial L}{\partial t}\frac{\partial^2 L}{\partial \dot{q}'^2} + \Lambda\frac{\partial^3 L}{\partial t \, \partial \dot{q}'^2} + \dot{q}'\frac{\partial \Lambda}{\partial q}\frac{\partial^2 L}{\partial \dot{q}'^2} + \dot{q}'\Lambda\frac{\partial^3 L}{\partial q \, \partial \dot{q}'^2}\right) d\dot{q}'$$

$$+ \frac{\partial A}{\partial t} - \frac{\partial B}{\partial q}. \tag{5.37}$$

Notice that the integrand is a function of $\dot{q}'$, the integration variable, and the only place that $\dot{q}$ appears in that side is in the upper limit of the integral. Into this equation we put the $d\Lambda/dt$ equation, Eq. (5.36), and find that the integral may be performed exactly. The result is

$$\frac{\partial A}{\partial t} - \frac{\partial B}{\partial q} = \Lambda(q, c, t)\left(\frac{\partial^2 L}{\partial q \, \partial \dot{q}}\dot{q} + \frac{\partial^2 L}{\partial t \, \partial \dot{q}} - \frac{\partial L}{\partial q}\right)_{\dot{q}=c}. \tag{5.38}$$

Thus the right side is a function only of q, t, and this equation is trivial to solve. (For example, see Exercise 5.8.) This completes the one-dimensional discussion.

5.3   Many Dimensions

The multidimensional case is significantly different from the one-dimensional theory. Only some results are known, and we here will outline some of the more interesting features. No proofs will be given.

Let $L(q^i, \dot{q}^i, t)$ be a Lagrangian. The equations of motion will be written $L_i = 0$, where

$$L_i = W_{ij}\, \ddot{q}^j + V_{ij}\, \dot{q}^j + U_i, \tag{5.39}$$

where for convenience we have defined

$$
\begin{aligned}
W_{ij} &= \frac{\partial^2 L}{\partial \dot{q}^i\, \partial \dot{q}^j}, \\[2mm]
V_{ij} &= \frac{\partial^2 L}{\partial \dot{q}^i\, \partial q^j}, \\[2mm]
U_i &= \frac{\partial^2 L}{\partial \dot{q}^i\, \partial t} - \frac{\partial L}{\partial q^i}.
\end{aligned}
\tag{5.40}
$$

When we discuss an equivalent Lagrangian $\bar{L}$, we will denote by $\bar{W}_{ij}$, and so on, the corresponding partial derivative functions. Call W^{ij} the inverse of W_{ij}, as defined by the equation

$$W^{is} W_{sj} = \delta^i{}_j. \tag{5.41}$$

(We assume that W_{ij} is nonsingular. The theory pertaining to a Lagrangian with a singular W_{ij} is interesting, also incomplete, and will not be discussed here.)

The equations of motion therefore can be written

$$\ddot{q}^i = W^{is}(-V_{st}\, \dot{q}^t - U_s). \tag{5.42}$$

If a second Lagrangian $\bar{L}$ exists, its equations of motion,

$$\ddot{q}^i = \bar{W}^{is}(-\bar{V}_{st}\, \dot{q}^t - \bar{U}_s) \tag{5.43}$$

must be identical to the first set. We define the fouling matrix Λ^i_j by

$$\Lambda^i_j = W^{is}\, \bar{W}_{sj}. \tag{5.44}$$

Therefore, we must have

$$\bar{V}_{jt}\, \dot{q}^t + \bar{U}_j = \Lambda^s_j (V_{st}\, \dot{q}^t - U_s), \tag{5.45}$$

where the equation holds identically, whether the equations of motion are satisfied or not. We call (5.45) the basic equation for Λ^i_j.

The first step in investigating the properties of Λ^i_j is to find its equation of motion $d\Lambda^i_j/dt = X^i{}_j$. The procedure makes use of the basic equation and its partial derivative with respect to $\dot{q}^k$. The result is not that $X^i{}_j$ is zero, but that the trace $X^s{}_s$ vanishes when the equations of motion are satisfied. Thus the trace of Λ^i_j, namely, Λ^s_s, is a constant of the motion. Similarly, it can be shown that the trace of any power, such as $\Lambda^s_t \Lambda^t_u \Lambda^u_s$, is a constant of the motion: Any eigenvalue of Λ^i_j is a constant of the motion.

The unsolved aspects of the problem have to do with whether any equivalent $\bar{L}$ exists. When, given a matrix Λ^i_j, all of whose eigenvalues are constants of the motion (the individual entries need not be), can an $\bar{L}$ be found so that $\bar{L}_j = \Lambda^i_j L_i$? Sometimes no second Lagrangian exists; sometimes many do; mostly nobody knows.

When we go to field theory, the Lagrangian is typically expressed as the integral over space of a Lagrangian density (field theory is discussed in a later chapter):

$$L = \int \mathcal{L}\,dv, \tag{5.46}$$

where dv is the volume element. Here the fouling matrix becomes an integral operator, and an equivalent Lagrangian $\bar{L}$ may very well be nonlocal in space:

$$\bar{L} = \int \int \mathcal{L}'\,dv\,dv' \tag{5.47}$$

(so that the second Lagrangian density is itself an integral over space of a density-density).

All of this is currently under study. In a trivial case, however, many examples of equivalent Lagrangians may be generated: If L is of the form of a sum of terms, each dependent on one of the q^i, an equivalent $\bar{L}$ can be made by taking any linear combination of these terms, with nonzero coefficients. Once this is done, a quantization theory may be applied. The results show that classically equivalent Lagrangians may be inequivalent quantum mechanically. The canonical procedure is to form the Hamiltonian (which will be discussed in a later chapter) and to replace q^i and the conjugate momenta p_i by operators. The Schrödinger equation is then formed and, hopefully, solved. If L and $\bar{L}$ are both t-independent, the process is straightforward, but drastically different results may be found!

One example that is very interesting is the case when $L = L_{(1)} + L_{(2)}$, with the two terms each being a simple-harmonic-oscillator Lagrangian. Choose as $\bar{L}$ the combination $\bar{L} = L_{(1)} - L_{(2)}$. The invariance group of L is compact; the invariance group of $\bar{L}$ is noncompact. The eigenvalues that describe the states of the L and the states of the $\bar{L}$ system therefore form completely different sets. One can argue that $\bar{L}$ is quite unnatural because what corresponds to "energy" is non-positive-definite. Nonetheless, if all that is known is that the classical system is a two-dimensional oscillator with equal spring constants in the two directions, either L or $\bar{L}$ may be used for its Lagrangian; and the quantum picture is unclear. The subject of equivalent Lagrangians is fascinating, although it does seem unnerving at first to recognize that the Lagrangian corresponding to a given classical behavior is nonunique in general.

EXERCISES

5.1. Prove (5.1) and the statement following (5.2). First, let $f(q^i, t)$ be an arbitrary function. Show that

$$\frac{d(\partial \dot{f}/\partial \dot{q}^i)}{dt} - \frac{\partial \dot{f}}{\partial q^i} = 0.$$

Now let $L(q^i, \dot{q}^i, t)$ be a Lagrangian, and let k and $f(q^i, t)$ be an arbitrary nonzero constant and function, respectively. Define $\bar{L}$ by

$$\bar{L}(q^i, \dot{q}^i, t) = kL + \dot{f}.$$

Show that $\bar{L}$ and L give equivalent equations of motion (more precisely, the Euler-Lagrange equations formed from $\bar{L}$ are proportional to those from L; therefore, the two sets of equations are exactly equivalent).

5.2. Show that the Lagrangian $(\delta_{ij}\dot{q}^i\dot{q}^j)^{1/2}$ gives a singular $\bar{W}_{ij}$ for the free particle. Show that the equations of motion, however, do imply straight-line motion, but with arbitrary acceleration and velocity.

5.3. Consider the free particle in one dimension, with coordinate $q(t)$. Show that $c_1 = \dot{q}$ and $c_2 = q - \dot{q}t$ are constants of the motion. Show that any constant of the motion is a function of c_1 and c_2. [*Hint:* Show that the change of variables $\{q, \dot{q}, t\} \rightarrow \{c_1, c_2, \tau\}$ defined by $c_1 = \dot{q}$, $c_2 = q - \dot{q}t$, $\tau = t$ has nonzero Jacobian, and therefore any function, whether a constant of the motion or not, can be expressed as $f(c_1, c_2, \tau)$. Then show that if f is a constant of the motion, $\partial f/\partial \tau = 0$.]

5.4. In the free-particle example, where $\Lambda = c_2$, if G is given arbitrarily and F is then determined, show that the combination $F\dot{q} + G$ is the total time derivative of some function $Z(q, t)$. [*Hint:* To solve $F\dot{q} + G = dZ/dt$ is to solve the pair of equations

$$\frac{\partial Z}{\partial q} = F \qquad \text{and} \qquad \frac{\partial Z}{\partial t} = G.$$

Show that the integrability condition for this pair of equations is satisfied and that therefore Z may be found.]

5.5. Suppose that for a free particle in one dimension, $\Lambda = \frac{1}{2}\dot{q}^2 = \frac{1}{2}c_1^2$. Find the most general $\bar{L}$ that yields the equation of motion $\Lambda\ddot{q} = 0$.

5.6. Show that if $\Lambda = 1$, there exists a function $f(q, t)$ such that $\bar{L} - L = \dot{f}$.
[*Hint:* from (5.24),

$$\frac{\partial^2 \bar{L}}{\partial \dot{q}^2} = \frac{\partial^2 L}{\partial \dot{q}^2},$$

so that

$$\bar{L} = L + A(q, t)\dot{q} + B(q, t).$$

Why do you then have to show that $\partial A/\partial t = \partial B/\partial q$?]

5.7. Carry through the calculations leading from (5.35) to (5.38) using for c a function of q and t, $c(q, t)$.

5.8. Call the right side of the A, B Eq. (5.38) $G(q, t)$. Show that two forms of the solution are

$$A(q, t) = \frac{\partial \Psi}{\partial q} + \int G \, dt, \qquad B = \frac{\partial \Psi}{\partial t}$$

and

$$A(q, t) = \frac{\partial \Phi}{\partial q}, \qquad B = \frac{\partial \Phi}{\partial t} - \int G \, dq,$$

where Ψ and Φ are arbitrary functions of q, t.

5.9. Suppose that a particle obeys the equation of motion

$$\ddot{q} = -\frac{dV}{dq} \qquad \text{where} \quad V = V(q) \text{ only.}$$

Find all Lagrangians that do not explicitly depend on t and that yield this equation of motion. Consider in particular the simple harmonic oscillator. Its equation of motion, assuming that the mass is unity, is

$$\ddot{q} = -k^2 q \qquad \text{where} \quad k = \text{const}.$$

One Lagrangian is $L = \frac{1}{2}\dot{q}^2 - \frac{1}{2}k^2 q^2$. One constant of the motion is $E = \frac{1}{2}\dot{q}^2 + \frac{1}{2}k^2 q^2$. Find the Lagrangian $\bar{L}$ that results from using E as the fouling function; this is a special case of the calculations leading to Exercise 5.8. Another pair of constants of the motion are

$$C_1 = \dot{q}\cos kt + kq \sin kt,$$

$$C_2 = \dot{q}\sin k - kq \cos kt.$$

Verify that C_1 and C_2 are constants of the motion, and show that E can be expressed as a function of C_1 and C_2. (In fact, any constant of the motion can be expressed as a function of C_1 and C_2.) Find the Lagrangians that result from using C_1 and from using C_2 as fouling functions. Here it is easiest to use the general method outlined as part of the proof of the last part of the one-dimensional discussion.

6

ROTATIONS AND SPINORS

A solid body contains a very large number ($\sim 10^{23}$) of mass points. Following the motion of each molecule is impossible in both classical mechanics and in quantum mechanics. However, we can make the approximation that the speed of sound in the body is effectively infinite. If this approximation holds, the body is assumed perfectly rigid. "Rigid" means that the distance between any two points in the body is fixed and constant. How justified is such an assumption of infinite sound speed? What is really required is that the time for sound waves to travel across the body be short compared to the time over which changes (such as changes in the forces applied to the body) occur. Environmental effects on the body should therefore be "slow and gentle." Whether a given situation satisfies this requirement depends, first, on the degree of accuracy we desire. Steel can be considered rigid for many everyday circumstances, but in other situations its elasticity and/or its ductibility become important.

The orientation and position of a rigid body may be specified by giving only six parameters. The specification goes as follows: Let us pick three non-collinear points, A, B, and C, in the body. Three of the coordinates can be given by giving the Cartesian coordinates of point A. Once these coordinates are given, we know that the point B lies on a sphere of radius equal to the A–B distance. The location of B may then be specified by giving the polar angles, θ, ϕ (defined as usual in terms of the Cartesian coordinates), which give the direction from A to B. This does not yet specify the orientation of the body because the specifications so far given are unchanged if the body rotates about the A–B line. But this can be fixed by giving one angle, ψ, which specifies the position of C about the A–B line. These six parameters specify the position and orientation of the body uniquely. It is sensible to take point A to be the center of mass of the body, especially if there are no net external forces, for then, as we saw in Chapter 4, we can ignore the center-of mass-motion and concentrate instead only on the orientation.

A **rotation** is a coordinate transformation in Euclidean 3-space, which transforms from one rectangular coordinate system to another without shifting the origin. Recall Eqs. (1.36)–(1.38); a rectangular coordinate system is one in which Pythagoras' theorem is explicitly true in its simplest form:

$$ds^2 = dx^2 + dy^2 + dz^2, \tag{6.1}$$

where x, y, z are rectangular coordinates and dx, dy, dz are here interpreted as infinitesimal coordinate differences. This means that a rectangular coordinate system is one in which the tensor

$$ds^2 = g_{ij}\, \mathbf{d}x^i \otimes \mathbf{d}x^j \tag{6.2}$$

has components

$$g_{ij} = \delta_{ij} = \operatorname{diag}(1,1,1). \tag{6.3}$$

Consider now an invertible coordinate transformation of the type

$$\bar{x}^i = \mathbf{A}^{\bar{\imath}}{}_j\, x^j, \tag{6.4}$$

where $\mathbf{A}$ is a 3×3 matrix of real constants. Our notation writes the new coordinates barred, as is the corresponding index on the array $\mathbf{A}^{\bar{\imath}}{}_j$.

Then

$$\mathbf{d}\bar{x}^i = \mathbf{A}^{\bar{\imath}}{}_j\, \mathbf{d}x^j \tag{6.5}$$

and

$$(\mathbf{A}^{-1})^j{}_{\bar{\imath}}\, \mathbf{d}\bar{x}^i = \mathbf{d}x^j. \tag{6.6}$$

Hence

$$ds^2 = \delta_{ij}\, \mathbf{d}x^i \otimes \mathbf{d}x^j \tag{6.7}$$

$$= \delta_{k\ell}\, (\mathbf{A}^{-1})^k{}_{\bar{\imath}}\, (\mathbf{A}^{-1})^\ell{}_{\bar{\jmath}}\, \mathbf{d}\bar{x}^i \otimes \mathbf{d}\bar{x}^j.$$

The requirement that $\mathbf{A}$ be a rotation is that the new coordinates also be rectangular, that is, that we also have

$$ds^2 = \delta_{ij}\, \mathbf{d}\bar{x}^i \otimes \mathbf{d}\bar{x}^j, \tag{6.8}$$

and hence

$$\delta_{ij} = \delta_{k\ell} \left(\mathbf{A}^{-1}\right)^k{}_{\bar{\imath}} \left(\mathbf{A}^{-1}\right)^\ell{}_{\bar{\jmath}}. \tag{6.9}$$

If we think of $\mathbf{A}^{-1}$ as a matrix, it is natural to think of the first index as a row label and the second as a column label. After the summation of the indices of the δ-symbol on the right is carried out, the result is

$$\delta_{ij} = [(\mathbf{A}^{-1})^T \mathbf{A}^{-1}]_{ij}, \tag{6.10}$$

where the symbol T denotes the transpose matrix. Thus—without regard to the upper or lower placement of indices—we have

$$\begin{aligned} I &= (\mathbf{A}^{-1})^T \mathbf{A}^{-1} \\ &= \mathbf{A}^{-1}(\mathbf{A}^{-1})^T. \end{aligned} \tag{6.11}$$

These equations state that $(\mathbf{A}^{-1})^T = (\mathbf{A}^{-1})^{-1} = \mathbf{A}$; such a matrix is called a real **orthogonal** matrix, and the condition that a matrix describe a rotation is that it be a position-independent orthogonal matrix.

We may now inquire as to the transformation of the basis vectors, the $\partial_i = \partial/\partial x^i$. We have

$$\begin{aligned} \frac{\partial}{\partial \bar{x}^i} &= \frac{\partial x^j}{\partial \bar{x}^i} \frac{\partial}{\partial x^j} \\ &= (\mathbf{A}^{-1})^j{}_i \frac{\partial}{\partial x^j}. \end{aligned} \tag{6.12}$$

This expression, when compared to the corresponding expression for the $\mathbf{d}\bar{x}^i$, yields a remarkable simplification for orthogonal transformations—one, however, that obscures the general distinction between basis vectors and basis 1-forms.

Equation (6.5) reads

$$\begin{pmatrix} \mathbf{d}\bar{x}^1 \\ \mathbf{d}\bar{x}^2 \\ \mathbf{d}\bar{x}^3 \end{pmatrix} = [\mathbf{A}] \begin{pmatrix} \mathbf{d}x^1 \\ \mathbf{d}x^2 \\ \mathbf{d}x^3 \end{pmatrix}, \tag{6.13}$$

when $\mathbf{A}$ is regarded as a matrix, and the operators $\mathbf{d}x^1$, $\mathbf{d}x^2$, $\mathbf{d}x^3$ and $\mathbf{d}\bar{x}^1$, $\mathbf{d}\bar{x}^2$, $\mathbf{d}\bar{x}^3$ are treated like elements of "column" matrices. Now Eq. (6.12) reads

$$\begin{pmatrix} \dfrac{\partial}{\partial \bar{x}^1} \\[2mm] \dfrac{\partial}{\partial \bar{x}^2} \\[2mm] \dfrac{\partial}{\partial \bar{x}^3} \end{pmatrix} = [\mathbf{A}^{-1}]^T \begin{pmatrix} \dfrac{\partial}{\partial x^1} \\[2mm] \dfrac{\partial}{\partial x^2} \\[2mm] \dfrac{\partial}{\partial x^3} \end{pmatrix}, \tag{6.14}$$

but the orthogonality of $\mathbf{A}$ means that $[\mathbf{A}^{-1}]^T = \mathbf{A}$. Hence read as a matrix equation, this relation between column matrices (in which the elements are the basis vectors $\partial/\partial x^j, \partial/\partial \bar{x}^j$) is exactly the same form as that relating the 1-form bases and is also the form

$$
\begin{pmatrix} \bar{x}^1 \\ \bar{x}^2 \\ \bar{x}^3 \end{pmatrix} = [\mathbf{A}] \begin{pmatrix} x^1 \\ x^2 \\ x^3 \end{pmatrix},
\tag{6.15}
$$

which relates the coordinates of a point, x^1, x^2, x^3 to the coordinates in the new frame. In all cases, the same matrix $[\mathbf{A}]$ effects the transformation. If we were considering the general position-dependent coordinate transformation instead of a rotation, none of the three transformations would be mutually equal.

We may directly express the rotation in terms of angles by considering the effect of two-dimensional rotations. We may start by calculating in the two-dimensional $\{x, y\}$-coordinate plane, but we extend this result trivially to rotations about the z-axis in 3-space, since such rotations leave the z coordinate unchanged.

Consider the pair of basis vectors $\mathbf{e}_1 = \partial/\partial x, \mathbf{e}_2 = \partial/\partial y$, which are associated with the rectangular coordinate system $\{x, y\}$. Suppose that we consider a rotated rectangular coordinate frame in which $\bar{\mathbf{e}}_1 = \partial/\partial \bar{x}, \bar{\mathbf{e}}_2 = \partial/\partial \bar{y}$. Suppose that the angle of rotation is ϕ, measured by the angle from $\mathbf{e}_1$ to $\bar{\mathbf{e}}_1$. Then, resolving the components of $\bar{\mathbf{e}}_i$ onto $\mathbf{e}_j$, we have

$$
\begin{aligned}
\bar{\mathbf{e}}_1 &= \mathbf{e}_1 \cos\phi + \mathbf{e}_1 \sin\phi, \\
\bar{\mathbf{e}}_2 &= -\mathbf{e}_1 \sin\phi + \mathbf{e}_1 \cos\phi,
\end{aligned}
\tag{6.16}
$$

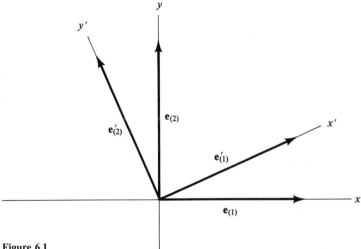

Figure 6.1

which is of the form of (6.14) with

$$
\mathbf{A} = \begin{pmatrix} \cos\phi & \sin\phi \\ -\sin\phi & \cos\phi \end{pmatrix},
\tag{6.17}
$$

which is an orthogonal matrix, as may be verified by direct calculation. The generalization to a three-dimensional rotation about the z-axis is straightforward:

$$\mathbf{A}_{3\times3} = \begin{pmatrix} \cos\phi & \sin\phi & 0 \\ -\sin\phi & \cos\phi & 0 \\ 0 & 0 & 1 \end{pmatrix}, \tag{6.18}$$

and rotations about the other axes can be obtained by relabeling the axes cyclically. The rotation just described has determinant $+1$; in fact, since $\det \mathbf{A}^T = \det \mathbf{A}$, we have

$$(\det \mathbf{A})\,(\det \mathbf{A}^T) = \det(\mathbf{A}\mathbf{A}^T), \tag{6.19}$$

so

$$(\det \mathbf{A})^2 = (\det \mathbf{I}) = 1. \tag{6.20}$$

We shall henceforth not consider those orthogonal matrices with $\det \mathbf{A} = -1$. [They can all be shown to be the product of a matrix with determinant $+1$ with an inversion, such as $\operatorname{diag}(-1,-1,-1)$.]

In the introductory section of this chapter we argued that a total of three angles suffice to specify a rotation; a fixed prescription for defining them is given in terms of **Euler angles**. Any orientation of a body is a rotation in a specified order, about three different axes, through three specified angles, ϕ, θ, ψ. With suitable specifications, the angles θ, ϕ, and ψ are the angles described in the introduction to this chapter.

The rotation matrices are:

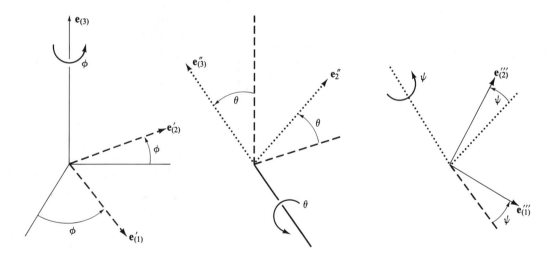

Figure 6.2

$\mathbf{A}_\phi$, a rotation by ϕ about $\mathbf{e}_3$, followed by   (6.21a)

$\mathbf{B}_\theta$, a rotation by θ about $\bar{\mathbf{e}}_1$, followed by   (6.21b)

$\mathbf{C}_\psi$, a rotation by ψ about $\bar{\bar{\mathbf{e}}}_3$,   (6.21c) where the barred basis is the result of **A** acting on the original rectangular basis system, and the double-barred basis results when **B** acts on the barred basis. The complete rotation matrix specified by the Euler angles is thus

$$\mathbf{R}(\phi, \theta, \psi) = \mathbf{C}_\psi \mathbf{B}_\theta \mathbf{A}_\phi. \tag{6.22}$$

R is clearly an orthogonal matrix because

$$\mathbf{R}^T \equiv \mathbf{A}^T \mathbf{B}^T \mathbf{C}^T = \mathbf{A}^{-1} \mathbf{B}^{-1} \mathbf{C}^{-1} = \mathbf{R}^{-1}. \tag{6.23}$$

The explicit **R** is constructed by matrix multiplication of its defining factors.

The rotations form a group (i.e., there is an identity rotation; the product of two rotations—that is, a rotation followed by a second rotation—is a rotation; and every rotation has an inverse). Orthogonal matrices whose determinant is +1—also known as **proper rotations**—are continuously connected to the identity. The meaning of *proper* can be best explained by an example of an *improper* rotation. Inverting three-dimensional space through the origin (i.e., $x \to -x$, $y \to -y$, $z \to -z$ in rectangular coordinates) is an orthogonal matrix of determinant -1; it is an improper rotation. All improper rotations can be written as the product of a proper rotation times an inversion of this type. "Small" rotations are those close to the identity, and a sequence of rotations can be specified, using, for instance, the Euler angles to give the continuous parametrization of proper rotations. The group, labeled SO(3,**R**) (unit-determinant [Special], Orthogonal, three-dimensional, Real matrices) is a compact 3-manifold.

It is instructive at this point to return to the case of two-dimensional rotations. It is well known that by introducing the complex numbers (instead of 2-vectors on the plane), rotations are represented by complex multiplication; rotation through an angle α becomes multiplication by $e^{i\alpha}$. The rotations, in the two-dimensional case, are specified by only one parameter; hence the unit-modulus object $e^{i\alpha}$ is used. We have not exerted the full power of the complex analysis because complex numbers contain, of course, two real numbers; the modulus of the rotation operator $e^{i\alpha}$ is constrained to unity.

In three-dimensional rotations, a single complex number does not contain the three real parameters that are required to specify the rotation. A set of two complex numbers contains one redundant coefficient. Nonetheless, we shall find a parameterization in terms of two complex numbers useful for describing rotations, once we have imposed a certain restriction, reminiscent of the restriction to unit-modulus complex numbers in the two-dimensional case above.

We begin by an analysis of two-dimensional complex vectors (called **spinors**)

$$\xi^A, \qquad A = 1, 2, \tag{6.24}$$

where each component ξ^A is a complex number. These spinors may be operated on by linear operators, such as 2×2 complex matrices:

$$\zeta^A = Q^A{}_B \xi^B. \tag{6.25}$$

The matrix $\mathbf{Q} = [Q^A{}_B]$ will represent rotations in three-dimensional space. To make the prescription precise, we demand that $\mathbf{Q}$ is **unitary** and that it have unit determinant. A unitary matrix is one that satisfies

$$\mathbf{Q}\mathbf{Q}^\dagger = \mathbf{Q}^\dagger\mathbf{Q} = \mathbf{I}, \qquad (6.26)$$

where the symbol $\dagger$ indicates the complex conjugate of the transpose of $\mathbf{Q}$. Taking the determinant of this equation gives

$$|\det \mathbf{Q}|^2 = 1; \qquad (6.27)$$

hence a further condition (we take $\det \mathbf{Q} = 1$) is required to fix the phase of $\mathbf{Q}$. Such matrices comprise the group of Special (i.e., unit determinant), two-dimensional Unitary matrices, abbreviated SU(2).

Suppose that we write out the components of $\mathbf{Q}$:

$$\mathbf{Q} = \begin{pmatrix} \alpha & \beta \\ \gamma & \delta \end{pmatrix}, \qquad (6.28)$$

where the entries are complex numbers. The unitary condition:

$$\begin{pmatrix} \alpha\alpha^* + \beta\beta^* & \alpha\gamma^* + \beta\delta^* \\ \gamma\alpha^* + \beta^*\delta & \gamma\gamma^* + \delta\delta^* \end{pmatrix} = \text{diag}\,(1,1) \qquad (6.29)$$

is a total of four conditions on the entries. In particular, the off-diagonal entries may be solved:

$$\alpha = -\frac{\beta\delta^*}{\gamma^*}. \qquad (6.30)$$

The additional constraint $\det \mathbf{Q} = 1$ gives

$$\alpha\delta - \beta\gamma = 1. \qquad (6.31)$$

Hence using (6.30) yields

$$-\frac{\beta}{\gamma^*(\delta^*\delta + \gamma\gamma^*)} = 1, \qquad (6.32)$$

or since $\delta\delta^* + \gamma\gamma^* = 1$ by (6.29),

$$-\beta = \gamma^*; \qquad (6.33)$$

but then $\delta = \alpha^*$. Consequently, our general SU(2) matrix is

$$\mathbf{Q} = \begin{pmatrix} \alpha & \beta \\ -\beta^* & \alpha^* \end{pmatrix}, \qquad (6.34)$$

which contains four real parameters, with the additional requirement that $\alpha\alpha^* + \beta\beta^* = 1$ [the so-far-unused component of (6.29)], a unit modulus requirement reminiscent of that of the two-dimensional case above.

Associated with every spinor ξ^A is its complex conjugate spinor $\bar{\xi}^{\bar{A}}$, where

$$\bar{\xi}^{\bar{1}} = (\xi^1)^*, \qquad (6.35)$$

$$\bar{\xi}^{\bar{2}} = (\xi^2)^*. \qquad (6.36)$$

If we have a transformation rule for spinors of the form

$$\eta^B = Q^B{}_A \xi^A,$$ (6.37)

the complex conjugate of this equation is

$$\bar{\eta}^{\bar{B}} = (Q^B{}_A)^* \bar{\xi}^{\bar{A}},$$ (6.38)

which shows that the complex conjugate of a spinor does not transform in the same way as the spinor does, but by the complex conjugate of the transformation matrix. The barred indices are introduced to take care of this distinction:

$$\bar{\eta}^{\bar{B}} = \bar{Q}^{\bar{B}}{}_{\bar{A}} \bar{\xi}^{\bar{A}},$$ (6.39)

where

$$\bar{Q}^{\bar{B}}{}_{\bar{A}} = (Q^B{}_A)^*.$$ (6.40)

Further, a general spinor that obeys the complex conjugate transformation law will be written $\gamma^{\bar{B}}$, the barred index indicating the conjugate transformation property. (This spinor is the complex conjugate of $\bar{\gamma}^B$, so the use of bars on the indices is not, as it at first seems, redundant.)

The SU(2) matrices form a group.

(a) There is the identity matrix $\bar{I}$.

(b) The product of two SU(2) matrices is an SU(2) matrix:

$$\det(\mathbf{UV}) = (\det \mathbf{U})(\det \mathbf{V}) = 1;$$

$$(\mathbf{UV})^\dagger (\mathbf{UV}) = \mathbf{V}^\dagger \mathbf{U}^\dagger \mathbf{UV} = \mathbf{I}.$$

(c) There is an inverse to every element:

$$\mathbf{U}^{-1} = \mathbf{U}^\dagger.$$

These three properties are unchanged under complex conjugation, so we have two copies of the group, usually denoted SU(2) and $\overline{\text{SU(2)}}$. The elements of $\overline{\text{SU(2)}}$ are distinguished by barred indices.

To relate the SU(2) transformation to 3-space rotations, let us make a particular linear association of a tensor $X^{A\bar{B}}$ in spin space with coordinate points x^i in 3-space. We define

$$X^{A\bar{B}} = x^i \boldsymbol{\sigma}_i{}^{A\bar{B}},$$ (6.41)

where the matrices $\boldsymbol{\sigma}_i = [\sigma_i{}^{A\bar{B}}]$ have definite expressions in rectangular coordinates and in a particular spin basis:

$$\sigma_x{}^{A\bar{B}} = \begin{pmatrix} 0 & 1 \\ 1 & 0 \end{pmatrix},$$ (6.42a)

$$\sigma_y{}^{A\bar{B}} = \begin{pmatrix} 0 & -i \\ i & 0 \end{pmatrix},$$ (6.42b)

$$\sigma_z{}^{A\bar{B}} = \begin{pmatrix} 1 & 0 \\ 0 & -1 \end{pmatrix}.$$ (6.42c)

The $\sigma_i{}^{A\bar{B}}$ transform as the components of a tensor in spin space and as the components of a 1-form in 3-space. The explicit matrices $\sigma_x, \sigma_y, \sigma_z$ are called **Pauli spin matrices**. In the frame we work in,

$$X^{A\bar{B}} = \begin{pmatrix} z & x - iy \\ x + iy & -z \end{pmatrix}. \tag{6.43}$$

The calculation of det X shows that

$$\det(X^{A\bar{B}}) = -x^2 - y^2 - z^2, \tag{6.44}$$

which is the negative of the length of the position vector in 3-space and is the object that is invariant under rotations. Now

$$\det(Q^D{}_A\,\bar{Q}^{\bar{C}}{}_{\bar{B}}\,X^{A\bar{B}}) = (\det \mathbf{Q})\,(\det \bar{\mathbf{Q}})\,(\det \mathbf{X})$$
$$= \det \mathbf{X}, \tag{6.45}$$

so that SU(2) transformations applied to $X^{A\bar{B}}$ provide the spinor version of three-dimensional active rotations.

Recovering the transformed x^i from $X^{A\bar{B}}$ may be done by inspection, or one may write

$$x^i = \tfrac{1}{2}\sigma^i{}_{A\bar{B}}\,X^{A\bar{B}}, \tag{6.46}$$

where the $\sigma^i{}_{A\bar{B}}$ have the same components as $\sigma_i{}^{A\bar{B}}$, in the fundamental spinor frame and the rectangular coordinate system we are using. The alert reader will notice that both in spin space and in 3-space there are two kinds of transformations available. First, we fix our bases, and they are modifiable only if we allow different representations for the $\sigma_i{}^{A\bar{B}}$ and $\sigma^i{}_{A\bar{B}}$. The actual rotation in 3-space is here viewed as an **active transformation**, and that is the meaning that must be assigned to phrases such as "evaluate the new vector in the fundamental frame in use." The fundamental frame is tied to the definitions of the σ's mentioned above.

We will now deduce that a rotation through ϕ about the z-axis is represented by a $\mathbf{Q}$ of the form

$$\mathbf{Q} = \exp\left(\frac{i}{2}\phi\,\boldsymbol{\sigma}_3\right), \tag{6.47}$$

where the exponential of a matrix is the matrix equal to

$$\exp \mathbf{B} = \mathbf{I} + \mathbf{B} + \frac{1}{2}\mathbf{B}^2 + \frac{1}{3!}\mathbf{B}^3 + \cdots. \tag{6.48}$$

Note that an element of the exponential of a matrix is not the exponential of the element unless the matrix is diagonal. The $\boldsymbol{\sigma}_i$ that appear in Eqs. (6.41) and (6.47) have special anticommutation and commutation properties:

$$\boldsymbol{\sigma}_i\boldsymbol{\sigma}_j + \boldsymbol{\sigma}_j\boldsymbol{\sigma}_i = 2\delta_{ij}\mathbf{I}, \tag{6.49a}$$

$$[\boldsymbol{\sigma}_i, \boldsymbol{\sigma}_j] = 2i\varepsilon_{ijk}\boldsymbol{\sigma}_k, \tag{6.49b}$$

where ε_{ijk} is the totally antisymmetric 3-index Levi-Civita symbol. Hence $\sigma_3{}^2 = \mathbf{I}$, and therefore

$$\mathbf{Q} = \exp\left(\frac{i}{2}\phi\boldsymbol{\sigma}_3\right) = \mathbf{I}\cos\frac{\phi}{2} + i\boldsymbol{\sigma}_3\sin\frac{\phi}{2}$$

$$= \begin{pmatrix} e^{i\phi/2} & 0 \\ 0 & e^{-i\phi/2} \end{pmatrix}. \tag{6.50}$$

By direct calculation we have, for $\mathbf{Q} = \exp\left(\frac{i}{2}\phi\boldsymbol{\sigma}_3\right)$,

$$Q^A{}_B \bar{Q}^{\bar{C}}{}_{\bar{D}} X^{B\bar{D}} = \begin{pmatrix} z & e^{i\phi}(x - iy) \\ e^{-i\phi}(x + iy) & z \end{pmatrix}. \tag{6.51}$$

Recalling the complex-algebraic treatment of rotations in the 2-plane, we see that this really is a rotation about the z-axis through an angle ϕ.

In a similar way we may consider the matrix

$$\mathbf{Q}' = \exp\left(i\frac{\lambda}{2}n^a\boldsymbol{\sigma}_a\right), \tag{6.52}$$

where n^a is a unit 3-vector and λ is an angle. Now, using (6.49a), we have

$$\mathbf{Q}' = \mathbf{I}\cos\frac{\lambda}{2} + i(n^a\boldsymbol{\sigma}_a)\sin\frac{\lambda}{2}$$

$$= \begin{pmatrix} \cos\frac{\lambda}{2} + in^3\sin\frac{\lambda}{2} & (in^1 + n^2)\sin\frac{\lambda}{2} \\ (in^1 - n^2)\sin\frac{\lambda}{2} & \cos\frac{\lambda}{2} + in^3\sin\frac{\lambda}{2} \end{pmatrix}. \tag{6.53}$$

By direct multiplication as in the previous example [and comparison with the SO(3,$\mathbf{R}$) form], we find that $\mathbf{Q}'$ is the SU(2) correspondent of a rotation through an angle λ about the axis n^a.

With the result above, we can construct the $\mathbf{Q}$ that corresponds to a particular Euler angle rotation:

$$\mathbf{Q}(\phi, \theta, \psi) = \exp\left(i\frac{\psi}{2}\boldsymbol{\sigma}_3\right)\exp\left(i\frac{\theta}{2}\boldsymbol{\sigma}_1\right)\exp\left(i\frac{\phi}{2}\boldsymbol{\sigma}_3\right)$$

$$= \begin{pmatrix} e^{i\psi/2} & 0 \\ 0 & e^{-i\psi/2} \end{pmatrix} \begin{pmatrix} \cos\frac{\theta}{2} & i\sin\frac{\theta}{2} \\ i\sin\frac{\theta}{2} & \cos\frac{\theta}{2} \end{pmatrix} \begin{pmatrix} e^{i\phi/2} & 0 \\ 0 & e^{-i\phi/2}. \end{pmatrix} \tag{6.54}$$

The fact that a 3-space vector is represented by a rank-2 spinor that transforms under the application of $\mathbf{Q}$ and $\mathbf{Q}^\dagger$ means that each appearance of $\mathbf{Q}$ involves only half-angles; when both $\mathbf{Q}$ and $\mathbf{Q}^\dagger$ act, trigonometric identities give results that involve the full angles ϕ, θ, ψ. This points out a curious fact. The SU(2) identity matrix $\mathbf{I}$ corresponds to the identity rotation (i.e., no rotation at all). But the SU(2) matrix $-\mathbf{I}$ also corresponds to the 3-space identity.

Hence, to every 3-space rotation matrix there are two SU(2) matrices. The spinor group SU(2) distinguishes a rotation through 4π from a rotation through 2π. (This "nonclassical 2-valuedness" is the important property of *spin* in quantum mechanical systems.) It is clear that for small angles the SO(3,$\mathbf{R}$) and SU(2) matrices can be put in one-to-one relationship, although we have seen that elements in SU(2) near $-\mathbf{I}$, as well as those near $\mathbf{I}$, map near the identity in SO(3,$\mathbf{R}$). We have seen that SU(2) matrices can be written as the exponential of $(i\lambda/2)n^a\sigma_a$. If we restrict our consideration to infinitesimal angles, we have $\mathbf{Q} \approx \mathbf{I} + (i\lambda/2)n^1\sigma_a$. The linear term $2\,d\mathbf{Q}/d\lambda|_{\lambda=0} = in^a\sigma_a$ is called an **infinitesimal generator** of the group. Recall that

$$[\sigma_i, \sigma_j] = 2i\varepsilon_{ijk}\sigma_k \qquad \text{(sum on } k\text{)}. \qquad (6.55)$$

The addition of a multiplication to a vector space (the collection of linear combinations of $\mathbf{I}$ and σ_i) gives what is called an **algebra**.

Instead of considering the generators of SU(2), we may consider the generators of SO(3,$\mathbf{R}$) rotations. We find that infinitesimal rotations about each of the 3-axes are of the form $\mathbf{I} + \phi\mathbf{M}_a$, where $\mathbf{I}$ is the 3×3 unit matrix, and

$$\mathbf{M}_x = \begin{pmatrix} 0 & 0 & 0 \\ 0 & 0 & 1 \\ 0 & -1 & 0 \end{pmatrix}, \qquad (6.56a)$$

$$\mathbf{M}_y = \begin{pmatrix} 0 & 0 & 1 \\ 0 & 0 & 0 \\ -1 & 0 & 0 \end{pmatrix}, \qquad (6.56b)$$

$$\mathbf{M}_z = \begin{pmatrix} 0 & 1 & 0 \\ -1 & 0 & 0 \\ 0 & 0 & 0 \end{pmatrix}. \qquad (6.56c)$$

Notice that

$$[\mathbf{M}_a, \mathbf{M}_b] = \varepsilon_{abc}\mathbf{M}_c \qquad \text{(sum on } c\text{)}, \qquad (6.57)$$

which means that the algebra of the generators of SO(3,$\mathbf{R}$) is the same as the algebra of generators of SU(2).

It is clear that finite rotations do not commute. On the other hand, infinitesimal rotations do commute to first order. This allows us linearly to add infinitesimal angles times the axis of rotation as ordinary vectors add in three-dimensional space. An immediate consequence of this fact is that angular velocities can be added like vectors.

To proceed, notice that the $\mathbf{M}_a$ above are given numerically by

$$(\mathbf{M}_a)_{bc} = \varepsilon_{abc}. \qquad (6.58)$$

Notice that the matrix product of two $\mathbf{M}_a$ is calculated by

$$(\mathbf{M}_a)_{bc}\,(\mathbf{M}_d)_{cf} = \delta_{af}\delta_{bd} - \delta_{ad}\delta_{bf}. \qquad (6.59)$$

If $a = d$, we find that

$$(\mathbf{M}_a{}^2)_{bf} = \delta_{af}\delta_{ba} - \delta_{bf} \qquad \text{(no sum on } a\text{)};$$

for instance,

$$(\mathbf{M}_1)^2 = \begin{pmatrix} 0 & 0 & 0 \\ 0 & -1 & 0 \\ 0 & 0 & -1 \end{pmatrix}. \qquad (6.60)$$

With these facts, one can explicitly calculate any rotation in the form

$$\mathbf{A}(\phi, \hat{\mathbf{n}}) = \exp(\phi n^a \mathbf{M}_a), \qquad (6.61)$$

where ϕ is the angle and $\hat{\mathbf{n}} = n^a \mathbf{e}_a$ is the unit vector defining the direction of the axis.

EXERCISES

6.1. The rotation group may be enlarged to include one of the symmetries of the Euclidean plane by appending the **translation group** T:

$$\bar{\mathbf{x}} = \mathbf{x} + c^a \mathbf{e}_a,$$

where the c^a are constants and the $\mathbf{e}_a \equiv \partial/\partial x^a$ are the unit basis vectors along the Euclidean axes. Clearly, successive translations commute. Note also that the $\mathbf{e}_a$ are the generators of the group of translations. By writing out the result of an infinitesimal translation followed by an infinitesimal rotation, and then in the opposite order, show that the commutations of these two procedures gives

$$\delta x^b = \bar{x}^{\bar{b}} - x^b = n^a \varepsilon c^i (\mathbf{M}_a)_{bi},$$

where the rotation was ε about the a-axis, and the translation was c^a. Hence the net effect is that of a translation in a rotated direction. The generators of the rotation and translation groups do not commute unless the translation is along the axis of the rotation.

6.2. Start with an object (e.g., a book) in some definite position. Apply a rotation $\mathbf{A}(\pi/2, \hat{\mathbf{n}}_1)$ followed by $\mathbf{A}(\pi/2, \hat{\mathbf{n}}_2)$ around two perpendicular axes.

(a) Demonstrate that the result depends on the order and in no case could it be regarded as the vector sum of the individual rotations.

(b) If $\mathbf{x}$ is the position vector of a point on a rigid body, which is then rotated by an angle θ around the origin, show that

$$\bar{\mathbf{x}} = (\hat{\mathbf{n}} \cdot \mathbf{x})\hat{\mathbf{n}} + \big(\mathbf{x} - \hat{\mathbf{n}}(\hat{\mathbf{n}} \cdot \mathbf{x})\big)\cos\theta + \hat{\mathbf{n}} \times \mathbf{x}\sin\theta.$$

Hence, obtain the product $\mathbf{A}(\pi/2, \hat{\mathbf{n}}_1)$ and $\mathbf{A}(\pi/2, \hat{\mathbf{n}}_2)$. Discuss this rotation and the one in the opposite order, with application to the experiment suggested in part (a). [*Hint:* The answer will involve a rotation by angle $\frac{2}{3}\pi$ in a direction $[1/\sqrt{3}](1, 1, 1)$; see V. Heine, *Group Theory in Quantum Mechanics,* Pergamon Press, Oxford, 1960.]

6.3. (a) If $\mathbf{B} = \mathbf{U}^\dagger \mathbf{A} \mathbf{U}$, where $\mathbf{U}$ is unitary, show that trace $\mathbf{B} = $ trace $\mathbf{A}$.

(b) If the complex vectors π_i are normalized and orthogonal:

$$\pi_i^* \cdot \pi_j = \delta_{ij},$$

and $\psi_i = A_{ij} \pi_j$ (sum on j) are also normalized and orthogonal, show that $\mathbf{A}$ is a unitary matrix.

6.4. Demonstrate that a rotation about a point whose radius vector $\mathbf{x}_0$ is not zero gives

$$\bar{\mathbf{x}} = \mathbf{A}\mathbf{x} - (\mathbf{A} - \mathbf{I})\mathbf{x}_0.$$

What are the transformation laws for the differential forms $\mathbf{d}\bar{x}^i$ and for the basis vectors $\partial/\partial \bar{x}^i$?

6.5. (a) Write out the explicit form for the general (three-dimensional) rotation matrix in terms of the Euler angles by carrying out the multiplication in Eq. (6.22).

(b) Write out the explicit form for a general SU(2) rotation matrix in terms of the Euler angles by carrying out the multiplication in Eq. (6.54).

6.6. A free particle is confined to the $x^3 = 0$ plane. Rewrite the Lagrangian in a form that uses coordinates uniformly rotating (angular speed ω) around the x^3-axis. Find the equations of motion for the particle, expressed in these coordinates.

7

DYNAMICS OF RIGID BODY MOTION

We have up to now considered only rotations through fixed angles. In this chapter we consider rotation as a dynamical feature of the motion. We define angular velocity as a three-dimensional vector by the following procedure. Suppose that

$$n^a \dot{\phi}(\mathbf{M}_a)_{bc} \, dt \tag{7.1}$$

is the incremental rotation through time dt, where n^a defines the axis of rotation. We write

$$\omega^i = \tfrac{1}{2}\varepsilon^{ibc}\dot{\phi}n^a(\mathbf{M}_a)_{bc} = \dot{\phi}n^i, \tag{7.2}$$

which defines a 3-vector based on the rank-2 generator $n^a(\mathbf{M}_a)_{bc}$. Notice that n^i can be a function of time.

We shall henceforth consider only the components of ω in right-handed Cartesian frames. Raising and lowering have no effect on indices in such frames, and we write most index positions down. (The angular velocity vector is actually a **pseudovector**. Under a coordinate inversion, the components of a vector, such as velocity, change sign. Angular momentum and angular velocity are formed from products of two objects with that type of transformation rule, and hence their components do not change sign under inversion. This is the characteristic defining a pseudovector. With our restriction to right-handed frames only, this distinction is unimportant.)

Let us consider the angular momentum

$$\mathbf{L} = \sum_a m_a (\mathbf{x} \times \mathbf{v})_a \tag{7.3}$$

of a collection of masses, labeled by a. We write $p_a{}^i = m_a v^i$ for the momentum of particle a and assume the collection of masses is a rigid system; then the velocity $\mathbf{v}$ is $\boldsymbol{\omega} \times \mathbf{x}$ as verified by using the right-hand rule to determine the direction, and the rule $|\boldsymbol{\omega} \times \mathbf{x}| = |\omega| |x| \sin \lambda$, where λ is the angle between the vectors.

Hence, since $v_\ell = (\boldsymbol{\omega} \times \mathbf{x})_\ell = \varepsilon_{\ell bc} \omega^b x^c$,

$$L_i = \sum_a m_a \, \varepsilon_{ibc} \, x^b \varepsilon_{cdf} \, x^f \omega^d. \tag{7.4}$$

In (7.4) we drop the mass label a on each appearance of the coordinates x^k. Note that the angular momentum is linear in $\boldsymbol{\omega}$. Now $\varepsilon_{abc} = \varepsilon_{cab}$ and

$$\varepsilon_{cab} \, \varepsilon_{cdf} = \delta_{ad} \, \delta_{bf} - \delta_{af} \, \delta_{bd}, \tag{7.5}$$

which gives

$$\begin{aligned} L_i &= \sum_a m_a (\delta_{id} \, \delta_{bf} - \delta_{if} \, \delta_{bd}) x^b x^f \omega^d \\ &= \sum_a m_a (x_b \, x^b \delta_{id} - x_i \, x_d) \omega^d. \end{aligned} \tag{7.6}$$

This last form may be written in 3-vector notation:

$$\mathbf{L} = \sum_a m_a [(\mathbf{x} \cdot \mathbf{x}) \boldsymbol{\omega} - \mathbf{x}(\mathbf{x} \cdot \boldsymbol{\omega})]. \tag{7.7}$$

An alternative way to write this expression is

$$L_i = I_{ib} \, \omega^b, \tag{7.8}$$

where I_{ij} is the **moment of inertia** tensor. This tensor may be determined as a sum from (7.6). In a continuous system, it is given as an integral:

$$I_{ij} = \int \rho \, dV (x_k \, x^k \delta_{ij} - x_i \, x_j), \tag{7.9}$$

where ρ is the density and dV is the volume element $dV = d^3x$.

I appears when we calculate the kinetic energy T of a rotating rigid body. The basic definition of T is

$$T = \tfrac{1}{2} \sum_a m_a \mathbf{v} \cdot \mathbf{v}.$$

It is convenient to substitute $\boldsymbol{\omega} \times \mathbf{x}$ for the second $\mathbf{v}$ only; a use of the triple identity results in

$$T = \tfrac{1}{2} \sum_a m_a \, \mathbf{v} \cdot (\boldsymbol{\omega} \times \mathbf{x})$$

$$= \tfrac{1}{2} \sum_a m_a \, \boldsymbol{\omega} \cdot (\mathbf{x} \times \mathbf{v}).$$

Hence, using the defining relation for the angular momenta, we have

$$T = \tfrac{1}{2} \, \boldsymbol{\omega} \cdot \mathbf{L} = \tfrac{1}{2} \sum_a \omega^i I_{ij} \, \omega^j. \tag{7.10}$$

The matrix I_{ij} is a symmetric 3×3 matrix. Such a matrix can always be diagonalized by a rotation. It is desirable to find the frame in which this matrix is diagonal, since the relationship between $\mathbf{L}$ and $\boldsymbol{\omega}$ is simple in that frame. We shall see that particular axes fixed in the body yield diagonal $\mathbf{I}$. If the body has a symmetry, these body axes are the symmetry axes.

The procedure of the diagonalization of $\mathbf{I}$ means that by a rotation we find a set of basis vectors $a_k^{(n)} \boldsymbol{\partial}_k$ that solve

$$I_{jk} \, a_k^{(n)} = \lambda^{(n)} a_j^{(n)} \qquad \text{(no sum on } n\text{)}. \tag{7.11}$$

Here the label (n) indicates which vector we are considering. This says that the effect of matrix multiplication by $\mathbf{I}$ on one of the basis vectors yields a vector proportional [by the factor $\lambda^{(n)}$] to the original vector. This is exactly what we expect when a diagonal matrix acts on one of the axis vectors. The factors $\lambda^{(1)}$, $\lambda^{(2)}$, $\lambda^{(3)}$ are called the **eigenvalues**, or proper values, of the matrix I_{ab}. They are the diagonal entries in the diagonalized matrix.

The eigenvalue equation (7.11) above is, in fact, three sets of three equations. Each set is labeled by a value of (n) and determined by one of three possible eigenvalues $\lambda^{(n)}$ that may be inserted. The free 3-vector index j then labels the individual equations in the set.

How does one go about solving Eq. (7.11)? A two-step process is required, since we know a priori neither the $\lambda^{(n)}$ nor the $a_j^{(n)}$. We begin by noting that Eq. (7.11), rewritten

$$(I_{ab} - \lambda^{(n)} \delta_{ab}) \, a_b^{(n)} = 0, \tag{7.12}$$

is a linear, singular equation for $a_b^{(n)}$. It is singular because it is not invertible; a presumably nonzero vector $\mathbf{a}^{(n)}$ is annihilated by the matrix whose components are in the parentheses of (7.12). To write this matrix, we distinguish between the moment of inertia matrix $\mathbf{I}$ and the 3×3 identity matrix 1. The matrix that annihilates $\mathbf{a}^{(n)}$ is then

$$(\mathbf{I} - \lambda^{(n)} 1)_{ab}. \tag{7.13}$$

The determinant of a singular matrix vanishes; the result is a cubic polynomial equation in the variable λ:

$$\det(\mathbf{I} - \lambda 1) = 0. \tag{7.14}$$

This equation is a necessary requirement that Eq. (7.11) be solvable. This requirement determines the three $\lambda^{(n)}$, which are the roots of this equation. Once we have solved this problem, we may pick one of the $\lambda^{(n)}$ and solve for the corresponding $\mathbf{a}^{(n)}$. If the value of λ is not chosen as one of the three $\lambda^{(n)}$, then $\mathbf{I} - \lambda 1$ is nonsingular (and thus has an inverse); then $\mathbf{a} = 0$.

Even when λ is chosen to be one of the roots $\lambda^{(n)}$, Eq. (7.12) shows that $\mathbf{a}^{(n)}$ is not completely determined. Because the equation is linear, if $\mathbf{a}^{(n)}$ solves the equation, so does $\alpha \mathbf{a}^{(n)}$, where α is a constant. Hence only the direction of $\mathbf{a}^{(n)}$ will be determined, and we are free to normalize the length of $\mathbf{a}^{(n)}$ as we desire.

We show that for a real symmetric matrix I_{ab}, the eigenvalues $\lambda^{(n)}$ are real. We write

$$\mathbf{I}\mathbf{a}^{(n)} = \lambda \mathbf{a}^{(n)}. \tag{7.15}$$

The transposed complex conjugate of this equation is

$$\bar{\mathbf{a}}^{T(n)} \mathbf{I} = \bar{\lambda}^{(n)} \bar{\mathbf{a}}^{T(n)}, \tag{7.16}$$

where the bar means complex conjugation and $\mathbf{a}^{(n)}$ is thought of as a column vector and its transpose as a row vector. Note that $\mathbf{I}$ is a real symmetric matrix, so $\bar{\mathbf{I}}^T = \mathbf{I}$. Now multiply Eq. (7.15) on the left by $\bar{\mathbf{a}}^{T(n)}$ and Eq. (7.16) on the right by $\mathbf{a}^{(n)}$ and subtract:

$$\bar{\mathbf{a}}^{T(n)} \mathbf{I} \mathbf{a}^{(n)} - \bar{\mathbf{a}}^{T(n)} \mathbf{I} \mathbf{a}^{(n)} = (\lambda^{(n)} - \bar{\lambda}^{(n)}) \bar{\mathbf{a}}^{T(n)} \mathbf{a}^{(n)}. \tag{7.17}$$

Since the left side of this equation vanishes but $\mathbf{a}^{T(n)} \mathbf{a}^{(n)} > 0$, we have $\lambda^{(n)} = \bar{\lambda}^{(n)}$; in other words, the eigenvalues $\lambda^{(n)}$ are real.

A similar exercise shows that if two eigenvalues are distinct, the eigenvectors they correspond to are orthogonal: Suppose that $\lambda^{(n)}$ is different from $\lambda^{(m)}$. The transpose of the eigenvalue equation for $a^{(m)}$ is

$$\mathbf{a}^{T(m)} \mathbf{I} = \lambda^{(m)} \mathbf{a}^{T(m)}. \tag{7.18}$$

Multiply Eq. (7.15) by $\mathbf{a}^{T(m)}$ on the left and multiply Eq. (7.18) by $\mathbf{a}^{(n)}$ on the right, and subtract:

$$0 = (\lambda^{(m)} - \lambda^{(n)}) \mathbf{a}^{T(m)} \mathbf{a}^{(n)}.$$

The product $\mathbf{a}^{T(m)} \mathbf{a}^{(n)}$ is just the Euclidean dot product of the two vectors. Hence we conclude that the vectors are orthogonal if the eigenvalues are different.

If the eigenvalues are not different, so that $\lambda^{(n)} = \lambda^{(m)} = \lambda$, we have

$$(\mathbf{I} - \lambda)\mathbf{a}^{(n)} = 0 \tag{7.19}$$

and

$$(\mathbf{I} - \lambda)\mathbf{a}^{(m)} = 0; \tag{7.20}$$

that is, two different eigenvectors satisfy the same equation. (Notice that if all three eigenvalues $\lambda^{(n)}$ are the same, then the diagonal form of I_{ab} is $\lambda\delta_{ab}$; since rotations leave δ_{ab} invariant, the moment of inertia is diagonal in any Cartesian frame in this case.) If exactly two eigenvalues are equal, we proceed by solving for one of the eigenvectors associated with the repeated eigenvalue λ; call it $\mathbf{a}$. Now $\mathbf{a}$ has components along the $\partial_i = \partial/\partial x^i$ basis vectors: $\mathbf{a} = a^i \partial_i$. We now perform a rotation so that $\mathbf{a}$ is parallel to the new 3-axis. Suppose that $\mathbf{a}$ is normalized: $\mathbf{g}(\mathbf{a},\mathbf{a}) = 1$, which means that $\delta_{ij}\, a^i a^j = 1$. Define $\phi \equiv \tan^{-1}(a^2/a^1)$ and $\theta = \cos^{-1}(a^3)$. Then the rotation $\mathbf{R} = \mathbf{R}_2(\phi)\mathbf{R}_3(\theta)$, with

$$\mathbf{R}_3(\phi) = \begin{pmatrix} \cos\phi & \sin\phi & 0 \\ -\sin\phi & \cos\phi & 0 \\ 0 & 0 & 1 \end{pmatrix}, \tag{7.21}$$

$$\mathbf{R}_2(\theta) = \begin{pmatrix} \cos\theta & 0 & -\sin\theta \\ 0 & 1 & 0 \\ \sin\theta & 0 & \cos\theta \end{pmatrix}, \tag{7.22}$$

rotates the reference frame so that $\mathbf{a}$ is now ∂_3. In this frame $\mathbf{I}$ is given by

$$I_{ab} = \begin{pmatrix} I_{11} & I_{12} & 0 \\ I_{21} & I_{22} & 0 \\ 0 & 0 & \lambda \end{pmatrix}. \tag{7.23}$$

If I_{12} is not fortuitously zero, we may diagonalize the 2×2 block by a procedure similar to the 3×3 process, first finding the eigenvalues by solving

$$\det|\mathbf{I}_{2\times 2} - \lambda 1_{2\times 2}| = 0. \tag{7.24}$$

We already know that λ will be one of the roots of the 2×2 equation (7.24) and that the other root will be different from λ. (Or else we would have had three equal roots and a diagonal matrix to start with.) We may then rotate in the (1–2)-plane to obtain a completely diagonal I_{ab}, with orthogonal eigenvectors, which, of course, may be normalized.

Whether the three eigenvalues are equal or not, once the three eigenvalues and associated eigenvectors are determined, we may rotate the coordinate system as indicated above to align the coordinate axes with the eigenvectors. Because the eigenvectors $\mathbf{a}^{(n)}$ can be chosen to be orthonormal, we have

$$I_{ij} = \sum_n \lambda^{(n)} a_i^{(n)} a_j^{(n)}, \tag{7.25}$$

as may be verified by calculating $I_{ij} a_j^{(p)} = \lambda^{(p)} a_i^{(p)}$. Hence

$$\begin{aligned} a_i^{(p)}\, a_j^{(q)} I_{ij} &= \sum_n \lambda_{(n)}\, a_i^{(n)} a_j^{(n)} a_i^{(p)} a_j^{(q)} \\ &= \sum_n \lambda_{(n)}\, \delta^{(n)(p)}\, \delta^{(n)(q)} = \mathrm{diag}\,(\lambda^{(1)}, \lambda^{(2)}, \lambda^{(3)}). \end{aligned} \tag{7.26}$$

This equation means that the 3×3 array $a_i^{(p)}$ obtained by writing the three orthonormal eigenvectors as row vectors, one below the other, provides a 3×3

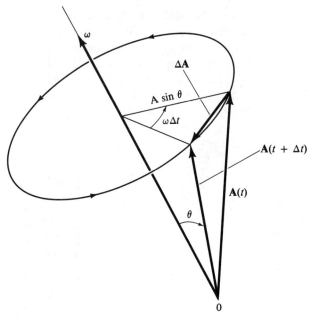

Figure 7.1

orthogonal matrix that defines the rotation that diagonalizes I_{ab}. Note that $a_i{}^{(p)}$ is an orthogonal matrix because $a_i{}^{(p)} a_i{}^{(q)} = \delta^{pq}$.

The coordinate axes that provide the diagonalized $\mathbf{I}$ are called the proper axes of $\mathbf{I}$. The proper axes of the moment of inertia are fixed in the body; we shall refer to a basis aligned with them as the **body frame**.

Let us continue to consider torque-free systems, so that

$$\frac{d\mathbf{L}}{dt} = 0. \tag{7.27}$$

In a rotating frame with instantaneous angular velocity $\boldsymbol{\omega}$ (Figure 7.1), a vector $\mathbf{A}$ that is fixed in the inertial frame experiences a time rate of change

$$\dot{\mathbf{A}}_{\text{rotating frame}} = -\boldsymbol{\omega} \times \mathbf{A}. \tag{7.28}$$

If the vector were also changing as viewed in the inertial frame, then

$$\dot{\mathbf{A}}_{\text{rotating frame}} = \dot{\mathbf{A}}_{\text{inertial frame}} - \boldsymbol{\omega} \times \mathbf{A}. \tag{7.29}$$

7.1 Equations of Motion in the Body Frame

Applying Eq. (7.28) to Eq. (7.27), we have

$$\dot{\mathbf{L}}_{\text{rotating frame}} + \boldsymbol{\omega} \times \mathbf{L} = 0, \tag{7.30}$$

where all components are projected into the rotating frame, as the equation of motion.

Hence, since the body frame (which is rotating) has been chosen to diagonalize the moment of inertia, we have

$$\frac{d(\lambda^{(i)}\omega_i)}{dt} + \sum_{k,\ell} \varepsilon_{ik\ell}\,\omega_k\,\lambda^{(\ell)}\omega_\ell = 0, \tag{7.31}$$

since $L_i = I_{ij}\,\omega^j$. Written out in components, this equation becomes (NO SUM is implied, even on repeated indices)

$$I_{xx}\dot\omega_x + \omega_z\omega_y(I_{zz} - I_{yy}) = 0, \tag{7.32a}$$

$$I_{yy}\dot\omega_y + \omega_x\omega_z(I_{xx} - I_{zz}) = 0, \tag{7.32b}$$

$$I_{zz}\dot\omega_z + \omega_y\omega_x(I_{yy} - I_{xx}) = 0, \tag{7.32c}$$

where we have introduced I_{xx}, I_{yy}, I_{zz}, for $\lambda^{(1)}, \lambda^{(2)}, \lambda^{(3)}$; they are the components of the diagonal moment of inertia in the body frame. Notice that if a torque acts on the body, then one introduces N_x, N_y, N_z, the components of the torque in the body frame, on the right side of these equations. Equations (7.32) are called the **Euler equations.**

A particularly simple example of torque-free motion arises when the body has a rotational symmetry axis. For instance, the Earth is an approximately axially symmetric body; it is approximately an oblate spheroid. Hence, if the origin is taken as the center of mass, $I_{xx} = I_{yy}$ for the Earth (where $\hat{\mathbf{z}}$ is the axis of symmetry). For such a body, the equations above have a straightforward solution: Immediately, one sees that

$$I_{zz}\dot\omega_z = 0; \tag{7.33}$$

that is, the z-component of the angular velocity is constant. Also,

$$I_{xx}\dot\omega_x + \omega_z\omega_y(I_{zz} - I_{xx}) = 0, \tag{7.34a}$$

$$I_{xx}\dot\omega_y + \omega_x\omega_z(I_{xx} - I_{zz}) = 0, \tag{7.34b}$$

where we used $I_{yy} = I_{xx}$. Now the quantity

$$\omega \equiv \frac{\omega_z(I_{zz} - I_{xx})}{I_{xx}} \tag{7.35}$$

is a constant (which is much smaller for the Earth than ω_z). Equations (7.34) are thus exactly like $\dot q$ and $\dot p$ for a harmonic oscillator. Differentiating the $\dot\omega_x$ equation (7.33a) gives

$$\ddot\omega_x + \omega\dot\omega_y = \ddot\omega_x + \omega^2\omega_x = 0, \tag{7.36}$$

which has the solution

$$\omega = \omega_0\cos\omega(t - t_0), \tag{7.37}$$

where ω_0 and t_0 are constants. Inserting this into the $\dot{\omega}_y$ equation (7.34b) gives

$$\omega_y = \omega_0 \sin \omega(t - t_0). \tag{7.38}$$

Hence ω_z, the projection of the angular velocity vector onto the symmetry axis, is always the same, but the point on the surface that is the axis of rotation moves slowly on a circle around the symmetry axis. For the Earth, which is oblate, $I_{zz} > I_{xx}$ and ω is in the same sense as ω_z. The magnitude is $\omega \sim 2\pi/300$ days, while $\omega_z \sim 2\pi/$day. The constant ω_0 is part of the initial data and could, in principle, be quite large. For the Earth, the pole wanders ~ 10 meters, so $\omega_0/\omega_z \sim 10\,\mathrm{m}/6 \times 10^6\,\mathrm{m} \sim 2 \times 10^{-6}$.

If torque $\mathbf{N}$ is nonzero, it enters on the right-hand sides of Eqs. (7.32). We can now give a simple example of motion when the system remains in a fixed orientation under the influence of torques: Consider a gyroscope suspended about two end supports along the axle. The Euler equations are

$$I_{xx}\dot{\omega}_x + \omega_z\omega_y(I_{zz} - I_{xx}) = N_x, \tag{7.39a}$$

$$I_{yy}\dot{\omega}_y + \omega_x\omega_z(I_{xx} - I_{zz}) = N_y, \tag{7.39b}$$

$$I_{zz}\dot{\omega}_z + \omega_y\omega_x(I_{yy} - I_{xx}) = N_z. \tag{7.39c}$$

In a perfectly balanced gyroscope, $N_i = 0$, but suppose there is a slight imbalance. An imbalance on one side of the gyroscope means that the principal moments are all unequal and that none of the principal axes point along the axle, but $\boldsymbol{\omega}$ is still constrained to lie along the axle by our hypothesis. Hence $\boldsymbol{\omega}$, when projected into the body frame, will have a constant projection along the principal direction that lies close to the axle, and a periodic component, which rotates in the orthogonal 2-plane of the body system. In other words, with $\omega_0 = \mathrm{const}$,

$$\omega_x = \omega_0 \sin\theta \sin\omega t,$$
$$\omega_y = \omega_0 \sin\theta \cos\omega t, \tag{7.40}$$
$$\omega_z = \omega_0 \cos\theta.$$

Inserting (7.40) into (7.39) gives

$$[I_{xx} + (I_{xx} - I_{zz})\cos\theta]\sin\theta\,\omega_0{}^2 \cos\omega t = N_x. \tag{7.41}$$

The torques do not vanish but instead have the instantaneous values necessary to keep the axle aligned with the supports. Because these torques are periodic, this amounts to shaking the bearings, a phenomenon that sometimes occurs in off-balance automobile wheels.

We have seen in Chapter 4 that the kinetic energy T splits into two parts, one that gives the kinetic energy as if the mass were all concentrated at a point and a second part that gives the kinetic energy in the center-of-mass frame. When dealing with a rigid body undergoing rotation, we write

$$T_{\mathrm{rot}} = \tfrac{1}{2}\omega_i\,\omega_j\,I_{ij} \tag{7.42}$$

for the energy measured in the center-of-mass frame (the origin must coincide with the center of mass). Here $\boldsymbol{\omega}_a$ is the instantaneous angular velocity vector. This vector can be expressed in any frame we wish. Two frames naturally suggest themselves. The first is the **space frame**, which is a nonrotating, inertial Cartesian frame but has its origin at the center of mass. Because the body is moving and rotating in this frame, the components of the moment of inertia, calculated according to Eq. (7.9), will be functions of time. On the other hand, we have already used the **body frame**, which is fixed in the body (hence is rotating and is not inertial). In this frame the integrals defining the components of the moment of inertia are constant, and it is reasonable to choose the body frame in which the axes are along the proper axes of the object. A calculation of the kinetic energy then requires that we project the angular velocity vector $\boldsymbol{\omega}_a$ into the body frame. Notice that this may be done at each instant, by instantaneously performing the projection (i.e., the fact that we are projecting into a rotating frame is irrelevant).

Because the angular velocity adds vectorially, the net angular velocity $\boldsymbol{\omega}$ can be expressed as a sum of the Euler angular velocity components $\boldsymbol{\omega}_\theta$, $\boldsymbol{\omega}_\phi$, $\boldsymbol{\omega}_\psi$. To perform operations based on the body axes, we need to understand how to project these vectors into the body frame. Note that $\boldsymbol{\omega}_\phi$ is a vector in the z-direction in the inertial frame. By projection into the rotated frame, we find $\boldsymbol{\omega}_\phi$ and also $\boldsymbol{\omega}_\theta$ and $\boldsymbol{\omega}_\psi$:

$$\boldsymbol{\omega}_\phi = \dot{\phi}\sin\theta\sin\psi\,\boldsymbol{\partial}_{x''} + \dot{\phi}\sin\theta\cos\psi\,\boldsymbol{\partial}_{y''} + \dot{\phi}\cos\theta\,\boldsymbol{\partial}_{z''}, \tag{7.43}$$

$$\boldsymbol{\omega}_\theta = \dot{\theta}\cos\psi\,\boldsymbol{\partial}_{x''} - \dot{\theta}\sin\psi\,\boldsymbol{\partial}_{y''}, \tag{7.44}$$

$$\boldsymbol{\omega}_\psi = \dot{\psi}\,\boldsymbol{\partial}_{z''}, \tag{7.45}$$

where double primes denote body axis coordinates. The results may be obtained directly by inspection from the figures or by use of the matrix (6.22) describing the Euler angle rotations.

Our general form for the kinetic energy expressed in the body frame is

$$T = \tfrac{1}{2}(I_{xx}\omega_x{}^2 + I_{yy}\omega_y{}^2 + I_{zz}\omega_z{}^2). \tag{7.46}$$

From Eqs. (7.43)–(7.45) we can read off the body frame angular velocity components $\omega_x, \omega_y, \omega_z$—written here without the double prime:

$$\omega_x = \dot{\phi}\sin\theta\sin\psi + \dot{\theta}\cos\psi,$$
$$\omega_y = \dot{\phi}\sin\theta\cos\psi + \dot{\theta}\sin\psi, \tag{7.47}$$
$$\omega_z = \dot{\phi}\cos\theta + \dot{\psi}.$$

Hence

$$T = \tfrac{1}{2}[I_{xx}(\dot{\phi}\sin\theta\sin\psi + \dot{\theta}\cos\psi)^2 + I_{yy}(\dot{\phi}\sin\theta\cos\psi - \dot{\theta}\sin\psi)^2$$
$$+ I_{zz}(\dot{\psi}\cos\theta + \dot{\psi})^2]. \tag{7.48}$$

Once we have this form, we can go ahead to obtain the potential energy for the solid body, write the Lagrangian, and obtain the equations of motion. In general, the integration of these equations requires the use of elliptic functions. Much has been written on the subject of gyroscopes and of tops; the quantum mechanical treatment is of great importance in nuclear physics.

7.2   Particle Dynamics in Rotating Frames

If we consider a 3×3 rotation matrix $\mathbf{O}$ as a time-dependent object, then it is sensible to consider its time derivative. We already know that infinitesimal rotations may be written

$$O_{ab} = \delta_{ab} + \phi n^c \varepsilon_{cab}, \qquad \phi \ll 1, \tag{7.49}$$

and the infinitesimal rotations commute. Now at any particular instant, $t + \varepsilon$, where ε is infinitesimal,

$$O_{ad}(t + \varepsilon) = (\delta_{ab} + \dot{\phi}\varepsilon n^c \varepsilon_{cab})O_{bd}(t) + O(\dot{\phi}\varepsilon)^2, \tag{7.50}$$

which simply says that the difference between the rotations measured at infinitesimally separated times is an infinitesimal rotation. Clearly, then, we find

$$\dot{O}_{ad} = \dot{\phi} n^c \varepsilon_{cab} O_{bd}(t), \tag{7.51}$$

or

$$\dot{O}_{ad}(O^{-1})_{df} = \omega_c \, \varepsilon_{caf}, \tag{7.52}$$

with

$$\omega_c = \dot{\phi} n^c, \tag{7.53}$$

which is the instantaneous angular velocity (pseudo-)vector. Notice that

$$\dot{O}_{ad}(O^{-1})_{df} = \dot{O}_{ad}O_{fd}$$

is antisymmetric and

$$\tfrac{1}{2}\varepsilon_{paf}\dot{O}_{ad}O_{fd} = \tfrac{1}{2}\omega_c \, \varepsilon_{caf} \, \varepsilon_{paf} = \tfrac{1}{2}(\delta_{ff}\,\delta_{cp} - \delta_{fp}\,\delta_{cf})\omega_c = \omega_p. \tag{7.54}$$

This result allows us to calculate the angular velocity from the derivative of the rotation, just as Eqs. (7.51)–(7.53) show how to compute the derivatives from ω_c.

Suppose we have two coordinate systems that are related to one another by a rotation. Then a particular point has coordinates in the two frames related by

$$x^i = O_{ij}\,\bar{x}^j. \tag{7.55}$$

The derivatives of the coordinates are related by

$$\dot{x}^i = \dot{O}_{ij}\,\bar{x}^j + O_{ij}\,\dot{\bar{x}}^j$$
$$\ddot{x}^i = \ddot{O}_{ij}\,\bar{x}^j + 2\dot{O}_{ij}\,\dot{\bar{x}}^j + O_{ij}\,\ddot{\bar{x}}^j. \tag{7.56}$$

Let us assume that the rotation has a fixed axis and a fixed angular velocity. Also, suppose that we evaluate the set of equations at $t = 0$ when $\mathbf{O}(t = 0) = 1$, a condition that can always be arranged by aligning the coordinate systems at the particular instant $t = 0$.

We must investigate the first and second derivatives of the infinitesimal rotation appearing in these equations. Since $\mathbf{O}(0) = 1$, we have

$$\mathbf{O}(t = \varepsilon) = \exp[-(\omega\varepsilon)n^a\varepsilon_{abc}], \tag{7.57}$$

where

$$\dot{\phi} = \omega \tag{7.58}$$

is the rotation angle during the interval $[0, \varepsilon]$. The minus sign in (7.57) is because of the form of (7.55); the barred frame is rotating at rate $\dot{\phi}$ around the axis n^a with respect to the unbarred (fixed) frame. Expanding the exponential, we find that

$$O_{bc}(\varepsilon) = \delta_{bc} - \varepsilon\omega n^a\varepsilon_{abc} + \tfrac{1}{2}(\varepsilon\omega)^2 n^a n^f\varepsilon_{abd}\,\varepsilon_{fdc} + \cdots. \tag{7.59}$$

In the limit, we have

$$\dot{O}_{bc}(t = 0) = \lim_{\varepsilon \to 0} \frac{1}{\varepsilon}[O_{bc}(\varepsilon) - \delta_{bc}] = -\omega n^a\varepsilon_{abc}, \tag{7.60}$$

which is just Eq. (7.51), the usual result for the angular velocity. In addition, the term of second order in ε yields

$$\ddot{O}_{bc}(t = 0) = (\omega)^2 n^a n^f\varepsilon_{abd}\,\varepsilon_{fdc}. \tag{7.61}$$

The results of the calculations above may be written

$$x^i = \bar{x}^i, \tag{7.62a}$$

$$\dot{x}^i = -\omega n^a\varepsilon_{aij}\,\bar{x}^j + \dot{\bar{x}}^i, \tag{7.62b}$$

$$\ddot{x}^i = (\omega)^2 n^a n^f\varepsilon_{aid}\,\varepsilon_{fdc}\,\bar{x}^c - 2\omega n^a\varepsilon_{aic}\,\dot{\bar{x}}^c + \ddot{\bar{x}}^i. \tag{7.62c}$$

The equations relating the velocities and accelerations in the two frames may be written in vector notation as

$$\mathbf{x} = \bar{\mathbf{x}}, \tag{7.63a}$$

$$\mathbf{v} = \dot{\mathbf{x}} = (\boldsymbol{\omega}\times\bar{\mathbf{x}}) + \dot{\bar{\mathbf{x}}}, \tag{7.63b}$$

$$\mathbf{a} = \dot{\mathbf{v}} = \ddot{\mathbf{x}} = (\mathbf{x}\times\boldsymbol{\omega})\times\boldsymbol{\omega} + 2(\boldsymbol{\omega}\times\dot{\bar{\mathbf{x}}}) + \ddot{\bar{\mathbf{x}}}, \tag{7.63c}$$

where our notation is that $\mathbf{x}$ is the position vector in an inertial frame and $\bar{\mathbf{x}}$ is the position vector in the rotating frame. Equation (7.63b) is a special case of

Eq. (7.29) relating components of the time derivative of a vector as expressed in relatively rotating frames.

In dealing with these equations we must remember that two assumptions have been made. The first is that $\boldsymbol{\omega}$ is a constant vector. The other is that the coordinate axes align at the time of interest $(t = 0)$. The component equations should have the additional rotation at the later time as a factor if the latter assumption is not made. This additional factor usually causes no problems; for instance, in Newton's equation, when $\mathbf{F}_{\text{inertial}}$ is the force in the inertial frame,

$$
\begin{aligned}
m^{-1}\mathbf{F}_{\text{inertial}} &= \mathbf{a}_{\text{inertial}} \\
&= \mathbf{a}_{\text{rotating}} + 2(\boldsymbol{\omega}\times\mathbf{v}_{\text{rotating}}) + \boldsymbol{\omega}\times(\boldsymbol{\omega}\times\mathbf{x}_{\text{rotating}}),
\end{aligned}
\tag{7.64}
$$

the force term can be taken into the rotating system only by a projection that correctly notes the position of the rotating frame basis. A constant external force then appears as a rotating force vector in the rotating frame. Important simplifications occur when either $\mathbf{F}_{\text{inertial}} = 0$, or when $\mathbf{F}_{\text{inertial}}$ is fixed in the rotating frame.

The "extra" terms that appear in the rotating frame are the **Coriolis force** $2(\boldsymbol{\omega}\times\mathbf{v})$ and the **centrifugal force** $\boldsymbol{\omega}\times(\boldsymbol{\omega}\times\mathbf{x})$. Because of the latter term, a force is necessary to remain at a fixed coordinate in the rotating system. The Coriolis force acts only upon objects moving in the rotating frame. These forces are often called "inertial forces," but we will not use this terminology to avoid confusion with expressions given in the inertial frame.

Consider a system fixed on the rotating earth at the equator, where the centrifugal and gravitational forces are (anti)parallel. A particle dropped from a height will undergo a net Coriolis force (seen in the rotating frame) of approximately $2m\dot{z}\omega$ eastward; this eastward force will give rise to motion that generates a small Coriolis term, which slightly modifies the z-direction motion but which we ignore. The principal effect of this eastward force is to make the dropped object fall ahead of the spot below its release point. This result is easily seen if we realize that at the top of the support, which is rotating with the earth, the tangential velocity (eastward) is greater than it is at the surface. Hence, the fall lands forward of the base of the support.

Another interesting case arises when one considers horizontal motion of a projectile emanating at $t = 0$ from the north pole, say. Here $2|\boldsymbol{\omega}\times\mathbf{v}| \sim 2\omega v$, and the Coriolis acceleration deviates the projectile westward. The angle of deviation is

$$
\theta \sim \frac{2\omega v(t^2/2)}{vt},
\tag{7.65}
$$

where the numerator is approximately the integrated displacement due to the westward force and the denominator is approximately the distance traveled. Hence

$$
\theta \sim \omega t,
\tag{7.66}
$$

the angle that the earth has turned under the projectile during its time of flight.

7.3   Rotating Frames and Larmor's Theorem

The Coriolis forces arising in rotating systems have a form like the magnetic force on a moving charge. This correspondence can be used to simplify problems involving charged particles moving in a magnetic field. To demonstrate this procedure, we first have to introduce the Lagrangian for motion in an external **electromagnetic field**:

$$L = \tfrac{1}{2} m \delta_{ij} \, \dot{x}^i \dot{x}^j + e \dot{x}^i A_i - e\phi, \tag{7.67}$$

where we have explicitly put in the metric of space in a Cartesian frame as δ_{ij} and where the vector potential A_i and the scalar potential ϕ are given functions of time and position; e is the electric charge. (This is the nonrelativistic Lagrangian.) The canonical momentum is

$$p_i = \frac{\partial L}{\partial \dot{x}^i} = m \delta_{ij} \, \dot{x}^j + e A_i, \tag{7.68}$$

which shows that the momentum contains terms that involve both the mechanical momentum and the electromagnetic vector potential. The equation of motion is

$$\left(\frac{\partial L}{\partial \dot{x}^i} \right)^{\cdot} = \frac{\partial L}{\partial x^i}, \tag{7.69}$$

which when written out becomes

$$m \delta_{ij} \, \ddot{x}^i + e \left(\frac{\partial A^i}{\partial t} + A_{i,k} \, \dot{x}^k \right) = e \dot{x}^p A_{p,i} - e \phi_{,i}. \tag{7.70}$$

By rearranging terms, we find that

$$m \delta_{ij} \, \ddot{x}^j = e \left(-\phi_{,i} - \frac{\partial A_i}{\partial t} \right) + e \dot{x}^p \left(A_{p,i} - A_{i,p} \right). \tag{7.71}$$

Since the electric field **E** and magnetic field **B** are

$$E_i = -\phi_{,i} - \frac{\partial A_i}{\partial t}, \qquad B_i = -\varepsilon_{ijk} \, A_{j,k}, \tag{7.72}$$

and since

$$\dot{x}^p \left(A_{p,i} - A_{i,p} \right) = -\dot{x}^p \, \varepsilon_{api} \, B_a = (\dot{\mathbf{x}} \times \mathbf{B})_i, \tag{7.73}$$

the equation of motion may be written

$$m \delta_{ij} \, \ddot{x}^j = e E_i + e (\dot{\mathbf{x}} \times \mathbf{B})_i, \tag{7.74}$$

which is the **Lorentz force** law. Note that the potentials A_i and ϕ admit gauge transformations: The forces arising from **E** and **B** are unchanged if

$$A_i \rightarrow A_i + \chi_{,i}; \qquad \phi \rightarrow \phi - \dot{\chi} \quad \text{for arbitrary scalar function } \chi(x^i, t).$$

Let us now consider a situation in which there is a uniform, time-independent **B**-field in the 3-direction and no electric field, so ϕ may be set equal to zero. Because of the gauge invariance, there are many **A**-potentials that give rise to such a field. One such is (where $B = $ const.)

$$A_i = \tfrac{1}{2}\varepsilon_{ijk}\, B_j x^k; \qquad B_j = B\delta_3{}^j. \tag{7.75}$$

Notice that

$$B_k = -\varepsilon_{k\ell s}\, A_{\ell,s} = -\varepsilon_{k\ell s}\, \tfrac{1}{2}\varepsilon_{\ell pq}\, x^q{}_{,s} B_p = -\tfrac{1}{2}\varepsilon_{k\ell s}\, \varepsilon_{\ell ps}\, B_q \tag{7.76}$$

does recover B_k as required.

Because of the similarity of the **B**-field terms to Coriolis forces, we can eliminate the effect of at least the dominant part of the magnetic field by going to a frame rotating about an axis parallel to **B**. We may rewrite the Lagrangian in terms of new coordinates via $x^j = O_{ji}\, \bar{x}^i$:

$$L = \tfrac{1}{2}m\delta_{ij}\,(O_{i\ell}\,\bar{x}^\ell)^\bullet\,(O_{jk}\,\bar{x}^k)^\bullet + e(O_{is}\,\bar{x}^s)^\bullet O_{ik}A_{\bar{k}}. \tag{7.77}$$

The terms proportional to m in the Lagrangian give rise to the Coriolis and centrifugal terms already discussed. The remaining term (proportional to e) is

$$e\dot{O}_{is}\,\bar{x}^s O_{ik}\, A_{\bar{k}} + e\dot{\bar{x}}^s O_{is} O_{ik} A_k. \tag{7.78}$$

The last term in (7.78) will give rise to forces exactly like those in the inertial frame, but the term involving $e\dot{\mathbf{O}}$ is an additional term arising from the rotation.

Performing the explicit derivatives, we obtain

$$\begin{aligned}
L &= \tfrac{1}{2}m(\dot{O}_{i\ell}\,x^\ell + O_{i\ell}\,\dot{x}^\ell)\,(\dot{O}_{ik}\,x^k + O_{ik}\,\dot{x}^k) + e(\dot{O}_{ik}\,x^k + O_{ik}\,\dot{x}^k)O_{is}A_s \\
&= \tfrac{1}{2}m\Big[\dot{x}^k\dot{x}^k + 2x^k\dot{O}_{ik}\,\dot{x}^i + 2\Big(\frac{e}{m}\Big)\dot{x}^k A_k \\
&\qquad + \dot{O}_{ik}\dot{O}_{ip}\,x^k x^p + 2\Big(\frac{e}{m}\Big)\dot{O}_{ik}\,x^k A_i\Big],
\end{aligned} \tag{7.79}$$

where we further assume that **O** is instantaneously the identity (the fixed and rotating frames are instantaneously aligned) and drop the bars. Recall that for our fixed 3-direction magnetic field, $A_i = \tfrac{1}{2}\varepsilon_{ijk}\, B_j x^k$; further, $\dot{O}_{ik} = \varepsilon_{iks}\,\omega_s$. Hence

$$\begin{aligned}
L = \tfrac{1}{2}m[x^k\dot{x}^k &+ 2\dot{x}^k\big(\varepsilon_{kps}\,\omega_p\,x^s + \frac{e}{2m}\varepsilon_{kps}\, B_p\, x^s\big) \\
&+ \varepsilon_{iks}\,\varepsilon_{ipq}\,x^k x^p \omega_s\,\omega_q + \frac{e}{m}\varepsilon_{iks}\,\omega_s\,x^k \varepsilon_{ipq}\, B_p\, x^q].
\end{aligned} \tag{7.80}$$

By choosing $\omega_s = -(e/2m)B_s$, the terms linear in $\dot{\mathbf{x}}$ drop out of the Lagrangian, and we are left only with terms that represent a modified centrifugal potential.

In vector notation, the equation of motion in the rotating frame is

$$\ddot{\mathbf{x}}_{\text{rot}} + 2\boldsymbol{\omega}\times\dot{\mathbf{x}}_{\text{rot}} + \boldsymbol{\omega}\times(\boldsymbol{\omega}\times\mathbf{x}) = \frac{e}{m}[\dot{\mathbf{x}}_{\text{rot}}\times\mathbf{B} + \boldsymbol{\omega}\times(\mathbf{B}\times\mathbf{x})]. \tag{7.81}$$

Clearly, if $\omega = -eB/2m$ (called the Larmor frequency), the terms first order in the angular velocity vanish. The remaining second-order terms do not exactly cancel but give rise as expected to a modified centrifugal force. Hence we have:

*Larmor's Theorem*_____

To first order, the dynamical effects of a constant magnetic field can be removed by going to a rotating frame; in other words, the first-order effects of a constant magnetic field are equivalent to motion observed in a rotating frame.

_____

For atomic systems where e/m is the same for a large number of objects, Larmor's theorem allows substantial conceptual and analytical simplification of motion in a magnetic field.

EXERCISES

7.1. In a principal-axis body frame, compute the moment of inertial tensor of a spherical shell; of a solid sphere. Take the center of mass as the origin and express your result in terms of the mass of the object.

7.2. Suppose that I know the moment of inertia tensor in some particular frame. Derive a formula giving this tensor with the origin located in a different place.

7.3. A rigid object consists of four equal mass spheres (mass m), held rigid by framework of negligible mass. The spheres are at

$$\mathbf{x}_1 = (0,0,0), \qquad \mathbf{x}_1 = (0,2,0), \qquad \mathbf{x}_3 = (1,0,0), \qquad \mathbf{x}_4 = (2,0,0).$$

[Note that they all lie in the $(x–y)$-plane.] Calculate the moment of inertia tensor for this object, about its center of mass, and obtain its principal moments of inertia. Write the explicit three-dimensional rotation that transforms from the coordinates given here to the principal frame.

7.4. From the form of Eq. (7.7) deduce that I_{ij} is a nonnegative matrix. This implies that the eigenvalues $\lambda^{(n)}$ of **I** are all nonnegative (why?). [*Note:* A nonnegative matrix A_{ij} is one for which

$$\xi^i \xi^j A_{ij} \geq 0 \qquad \text{for all } \xi^i.]$$

7.5. O. Calame and J. D. Mulholland (*Science* **199**, 875, 1978) discuss the possibility that the observed free librations of the moon could have been excited by a meteor impact on the moon that was apparently observed from Earth in A.D. 1178. Suppose, as they did, that an object impacted on the moon at $\sim 20\,\text{km/s}$ and had a typical asteroid mass of $\sim 10^{17}\,\text{g}$. Assume also that it struck at some substantial fraction of the radius of the moon from the center and assume that the moon is axially symmetric but not spherical. Allen (*Astrophysical Quantities*, 3rd ed., Athlone Press, London, 1963) gives the average moment of inertia of the moon as $\sim 0.4\, M_{\text{moon}} b_{\text{moon}}^2$, with $M_{\text{moon}} \sim 7 \times 10^{25}\,\text{g}$ and $b_{\text{moon}} \sim 1.7 \times 10^6\,\text{m}$. How great would you expect the amplitude of the precession of the angular momentum induced by such a typical collision to be? (Calame and Mulholland give an "observed" result of $\sim 15\,\mathrm{m}$ at the surface of the moon for the amplitude of the motion.)

7.6. (a) Using the equation for the kinetic energy, Eq. (7.49), write down equations of motion for a torque-free rotating body. What are the constants of the motion?

(b) If one considers a top with a gravitational force acting on it, one may simply add to the Lagrangian the potential energy term $-V = -Mg\ell\cos\theta$, where ℓ is the distance from the support to the center of mass. Write the equations of motion for this situation. Which constants of the motion remain?

(c) Rewrite the results of (a) and (b) in the case of a symmetric rotator with $I_{xx} = I_{yy}$.

7.7. A thin-walled pipe of radius R, mass m, rolls without slipping on an inclined plane, which is inclined at an angle α to the horizontal. The only external force acting on the pipe is gravitation.

(a) What is the configuration space for this system? What is the state space?

(b) Write the Lagrangian for this system and obtain Lagrange's equations of motion.

(c) Solve the equations of motion for this system.

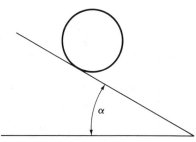

Figure P7.7

7.8. A rigid body consists of two uniform thin rectangular plates of width s and length $2s$ located in the (x^1-x^2)-plane, as illustrated. Regarding the coordinate system in the figure as body coordinates, compute the moments of inertia. Let M be the total mass of the system. Find the eigenvalues and eigenvectors of the moment of inertia tensor. Find a set of body coordinates in which the moment of inertia tensor is diagonal.

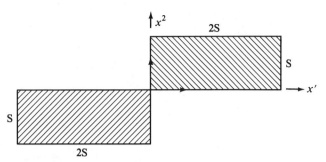

Figure P7.8

7.9. Find the principal moments of inertia and the principal axes (about the center of mass) of an object that consists of two infinitesimally thin rings (each of mass m) joined so that they lie in planes perpendicular to one another.

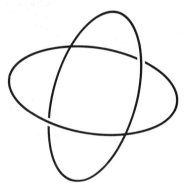

Figure P7.9

7.10. A hypothetical object consists of two rings, each of mass μ, radius R, which are joined by a rigid massless rod. The centers of the two rings are a distance d apart, and the rings lie in parallel planes that are orthogonal to the line joining the center of the rings. See Figure P7.10, p.116.

(a) Compute the moment of inertia of this object in any convenient frame (specify the frame the calculation is carried out in).

(b) Suppose that at $t = 0$ the angular velocity $\boldsymbol{\omega}$ of the object lies at an angle α with respect to the line joining the two rings. What is the angular momentum at that instant? What is the angular momentum and what is the value of the angle at time $t = \pi/\omega$. (No external forces or torques act.)

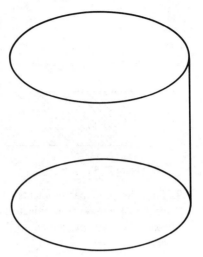

Figure P7.10

7.11. Determine the principal moments of inertia of a sphere of radius R, inside of which there is a spherical cavity of radius r whose center is offset a distance a from the center of the large sphere.

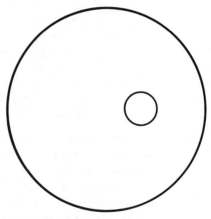

Figure P7.11

7.12. A common suggestion in designing satellites is that internal "gravity" be produced by rotating the structure so that the centrifugal force in the rotating frame holds objects against the hull.

    (a) Compute the angular velocity and the tangential speed for an internal environment of a particular gravity (e.g., $1\,\mathrm{g} \approx 10\,\mathrm{m/s^2}$), if the radius of the cylinder is $5\,\mathrm{m}$.

    (b) Compute the tension in the structure as a function of its mass density, angular velocity, and radius. Relate this to the internal gravity of part (a). By taking into account the fact that the tensile strength is typically proportional to the cross-sectional area of material, is it preferable to design a large or a small cylinder to obtain a particular value of g?

(c) Besides the centrifugal force inside the cylinder, one should also consider possible Coriolis forces and the effects of gradients of the centrifugal force. Make estimates for acceptable levels of these quantities and comment on the resultant restrictions on the design.

7.13. A free particle is confined to the $x^3 = 0$-plane. Rewrite the Lagrangian in coordinates that are uniformly rotating (angular speed ω) around the x^3-axis. Find the equations of motion for the particle in these coordinates.

7.14. The problem of an object dropped from the top of a tower, and that of the projectile fired from a gun at the North Pole on a spherical rotating Earth, can (neglecting atmospheric effects) be carried out in an inertial frame using the orbital techniques of Chapter 4. Do this analysis and compare the results with those obtained in this chapter.

7.15. Our discussion of Larmor's theorem showed that the lowest-order effects of a magnetic field are equivalent to observing the motion in a rotating frame. How is the centrifugal force modified, compared to simple motion from a rotating frame?

7.16. (a) Find the position, as seen by an observer on Earth, of a satellite in a circular orbit passing above the poles. State the position vector in a coordinate system attached to the surface of the Earth and in one with origin at the center of the Earth but rotating with the Earth.

(b) A hovering satellite is one placed in a circular orbit above a point on the equator and given a velocity just right to keep it exactly over that point constantly. Suppose that a satellite were placed in a circular orbit over a point on the equator with the correct speed (magnitude of velocity) to hover but at a slightly wrong angle. Describe its motion as seen from Earth.

8

HAMILTONIAN SYSTEMS

The Euler-Lagrange equations, for simple mechanical systems, are second-order, ordinary differential equations in the time. It is often much more useful to have a first-order system of equations to replace the second-order Euler-Lagrange equations. The most elegant procedure for generating the first-order system is to apply a **Legendre transformation** to the Lagrangian, transforming to a system where the independent variables are the coordinates and the canonical momenta. (The momenta are conserved when the Lagrangian does not contain the conjugate coordinates; hence the momenta are desirable dynamical variables.) Recall the definition of p_i, the canonical momentum conjugate to the coordinate q^i:

$$p_i = \frac{\partial L}{\partial \dot{q}^i}. \tag{8.1}$$

The **Hamiltonian** is the function $H(q^i, p_\ell, t)$ defined by

$$H(q^i, p_\ell, t) = \dot{q}^i \frac{\partial L}{\partial \dot{q}^i} - L, \tag{8.2}$$

via the procedure of eliminating appearances of $\dot{q}^i$ on the right side of (8.2) by using the relation (8.1).

8.1 Hamilton's Equations

Hamilton's equations are a set of first-order differential equations, which are equivalent to the second-order Euler-Lagrange equations. The latter involve the n variables q^i. Hamilton's equations are $2n$ equations in the $2n$ variables q^i, p_j. To derive Hamilton's equations, we first use Eq. (8.1) as an equation that may be solved to express $\dot{q}^i$ as a function of p_j, q^i, and t. This inversion is possible provided that $\partial p_j/\partial \dot{q}^i$ is a nonsingular matrix—that is, if $(\partial^2 L/\partial \dot{q}^i \partial \dot{q}^j)$ has nonzero determinant. (In case $L = T - V$ with V independent of $\dot{q}^i$ and if $T = \frac{1}{2}T_{ij}\,\dot{q}^i\dot{q}^j$, all that is required is that (T_{ij}) be nonsingular.) We then express H not as a function of q^i, $\dot{q}^i$, and t but as a function of q^i, p_i, and t.

The relation between H and L, given that $L = L(q, \dot{q}, t)$ while $H = H(q, p, t)$, is known as a *Legendre transformation*. We will now derive the equations that follow from such a Legendre transformation. We first give a succinct, if not very explicit, derivation, which we will repeat more carefully below. Demand that

$$
\begin{aligned}
dL &= \frac{\partial L}{\partial q}\,dq + \frac{\partial L}{\partial \dot{q}}\,d\dot{q} + \frac{\partial L}{\partial t}\,dt \\[2mm]
&= d(p\dot{q} - H) \\[2mm]
&= \dot{q}\,dp + p\,d\dot{q} - \frac{\partial H}{\partial q}\,dq - \frac{\partial H}{\partial p}\,dp - \frac{\partial H}{\partial t}\,dt.
\end{aligned}
\tag{8.3}
$$

The explicit quantities that H and L depend on specify the overall coefficients of dq, $d\dot{q}$, dp, and dt. Demanding the veracity of (8.3) thus leads to

$$
\frac{\partial L}{\partial \dot{q}} = p,
\tag{8.4a}
$$

$$
\frac{\partial L}{\partial q} = -\frac{\partial H}{\partial q},
\tag{8.4b}
$$

$$
\dot{q} = \frac{\partial H}{\partial p},
\tag{8.4c}
$$

$$
-\frac{\partial L}{\partial t} = \frac{\partial H}{\partial t}.
\tag{8.4d}
$$

Equation (8.4a) is just the definition of the momentum as given above. If we demand that Lagrange's equations are satisfied, we then find from (8.4b), since $\partial L/\partial q = dp/dt$,

$$
\dot{p} = -\frac{\partial H}{\partial q}.
\tag{8.5}
$$

Further, with (8.4c) and (8.5), we have from

$$\frac{dH}{dt} = \frac{\partial H}{\partial p}\dot{p} + \frac{\partial H}{\partial q}\dot{q} + \frac{\partial H}{\partial t}$$

a cancellation of all except the explicit time dependence:

$$\frac{dH}{dt} = \frac{\partial H}{\partial t}.$$

Collecting them together, Hamilton's equations are thus

$$\dot{p} = -\frac{\partial H}{\partial q},$$

$$\dot{q} = \frac{\partial H}{\partial p}, \qquad (8.6)$$

$$\dot{H} = \frac{\partial H}{\partial t}.$$

We will repeat the analysis leading to Eqs. (8.6), taking a more pedantic, explicit route. To be especially careful in setting up the equations, remember that a notation such as $\partial/\partial q^i$ is incomplete, for it does not say which variables are to be kept fixed. We therefore temporarily adopt the following convention: Let q^i, $\dot{q}^i$, and t, as usual, stand for the variables on which L depends. Let us call the variables on which H is to depend r^i, p_i, and τ, where

$$r^i = q^i, \qquad (8.7a)$$

$$p_i = \frac{\partial L}{\partial \dot{q}^i}, \qquad (8.7b)$$

$$\tau = t. \qquad (8.7c)$$

The reason for the last notation, Eq. (8.7c), is that $\partial/\partial\tau$ will mean that r^i and p_i are fixed, whereas $\partial/\partial t$ means that q^i and $\dot{q}^i$ are fixed. If L is explicitly time dependent, these two operations could be different. The notation $\partial/\partial r^i$ then will mean that the rest of the r^i's, the p_i, and τ are held fixed. The operation $\partial/\partial q^i$ means that the other q^i's, the $\dot{q}^i$, and t are held fixed. The notations $\partial/\partial p_i$ and $\partial/\partial\dot{q}^i$ have their appropriate definitions in their own systems.

Remember that we express

$$H = H(r, p, t) \qquad \text{and} \qquad L = L(q, \dot{q}, t), \qquad (8.8)$$

where

$$H = p_i \dot{q}^i - L \qquad \text{or} \qquad L = p_i \dot{q}^i - H, \qquad (8.9)$$

(with the necessary coordinate transformation, namely, Eqs. (8.7), being assumed). First, we compute the various elementary derivative operations (or vectors) as expressed by this change of basis:

$$\frac{\partial}{\partial q^i} = \frac{\partial r^s}{\partial q^i}\frac{\partial}{\partial r^s} + \frac{\partial p_s}{\partial q^i}\frac{\partial}{\partial p_s} + \frac{\partial \tau}{\partial q^i}\frac{\partial}{\partial \tau} = \frac{\partial}{\partial r^i} + \frac{\partial^2 L}{\partial q^i \partial \dot{q}^s}\frac{\partial}{\partial p_s}, \qquad (8.10a)$$

$$\frac{\partial}{\partial \dot{q}^i} = \frac{\partial r^s}{\partial \dot{q}^i}\frac{\partial}{\partial r^s} + \frac{\partial p_s}{\partial \dot{q}^i}\frac{\partial}{\partial p_s} + \frac{\partial \tau}{\partial \dot{q}^i}\frac{\partial}{\partial \tau} = \frac{\partial^2 L}{\partial \dot{q}^i\, \partial \dot{q}^s}\frac{\partial}{\partial p_s}, \tag{8.10b}$$

$$\frac{\partial}{\partial t} = \frac{\partial r^s}{\partial t}\frac{\partial}{\partial r^s} + \frac{\partial p_s}{\partial t}\frac{\partial}{\partial p_s} + \frac{\partial \tau}{\partial t}\frac{\partial}{\partial \tau} = \frac{\partial^2 L}{\partial t\, \partial \dot{q}^s}\frac{\partial}{\partial p_s} + \frac{\partial}{\partial \tau}. \tag{8.10c}$$

The next step is to apply these operations to $L = p_i\,\dot{q}^i - H$:

$$\frac{\partial L}{\partial q^i} = \dot{q}^s\frac{\partial p_s}{\partial q^i} - \frac{\partial H}{\partial q^i}$$

$$= \dot{q}^s\left(\frac{\partial p_s}{\partial r^i} + \frac{\partial^2 L}{\partial q^i\,\partial \dot{q}^t}\frac{\partial p_s}{\partial p_t}\right)$$

$$- \left(\frac{\partial H}{\partial r^i} - \frac{\partial^2 L}{\partial q^i\,\partial \dot{q}^t}\frac{\partial H}{\partial p_t}\right)$$

$$= \dot{q}^s\frac{\partial^2 L}{\partial q^i\,\partial \dot{q}^s} - \frac{\partial H}{\partial r^i} - \frac{\partial^2 L}{\partial q^i\,\partial \dot{q}^t}\frac{\partial H}{\partial p_t}, \tag{8.11}$$

$$\frac{\partial L}{\partial \dot{q}^i} = p_i + \dot{q}^s\frac{\partial p_s}{\partial \dot{q}^i} - \frac{\partial H}{\partial \dot{q}^i}$$

$$= p_i + \dot{q}^s\frac{\partial^2 L}{\partial \dot{q}^i\,\partial \dot{q}^t}\frac{\partial p_s}{\partial p_t} - \frac{\partial^2 L}{\partial \dot{q}^i\,\partial \dot{q}^s}\frac{\partial H}{\partial p_s}$$

$$= p_i + \dot{q}^s\frac{\partial^2 L}{\partial \dot{q}^i\,\partial \dot{q}^s} - \frac{\partial^2 L}{\partial \dot{q}^i\,\partial \dot{q}^s}\frac{\partial H}{\partial p_s}, \tag{8.12}$$

$$\frac{\partial L}{\partial t} = \dot{q}^s\frac{\partial p_s}{\partial t} - \frac{\partial H}{\partial t}$$

$$= \dot{q}^s\frac{\partial^2 L}{\partial t\,\partial \dot{q}^t}\frac{\partial p_s}{\partial p_t} - \frac{\partial^2 L}{\partial t\,\partial \dot{q}^s}\frac{\partial H}{\partial p_s} - \frac{\partial H}{\partial \tau}$$

$$= \dot{q}^s\frac{\partial^2 L}{\partial t\,\partial \dot{q}^s} - \frac{\partial^2 L}{\partial t\,\partial \dot{q}^s}\frac{\partial H}{\partial p_s} - \frac{\partial H}{\partial \tau}. \tag{8.13}$$

We now use the Euler-Lagrange equations and the definition of p_i. First, $\partial L/\partial \dot{q}^i = p_i$ and the stipulation that $\partial^2 L/\partial \dot{q}^i\partial \dot{q}^s$ be a nonsingular matrix imply that Eq. (8.12) for $\partial L/\partial \dot{q}^i$ reduces to

$$\frac{\partial H}{\partial p_s} = \dot{q}^s. \tag{8.14}$$

When Eq. (8.14) is substituted in Eq. (8.11) for $\partial L/\partial q^i$, we have

$$\frac{\partial L}{\partial q^i} = -\frac{\partial H}{\partial r^i}. \tag{8.15}$$

According to the Euler-Lagrange equations, $\partial L/\partial q^i = d(\partial L/\partial \dot{q}^i)/dt = \dot{p}_i$, so that

$$\frac{\partial H}{\partial r^i} = -\dot{p}_i. \tag{8.16}$$

Finally, it is seen that Eq. (8.13) for $\partial L/\partial t$ reduces to

$$\frac{\partial L}{\partial t} = -\frac{\partial H}{\partial \tau}. \tag{8.17}$$

It is usual to denote the variables on which H depends as q^i, p_i, and t. In that case $\partial/\partial q^i$ when applied to H really means $\partial/\partial r^i$. The operation $\partial/\partial t$ when applied to H means $\partial/\partial \tau$. Hamilton's equations are the equations for $\dot{q}^i$, $\dot{p}_i$, and $\partial H/\partial \tau$. The first two sets of equations are

$$\frac{\partial H}{\partial p_i} = \dot{q}^i, \tag{8.18}$$

$$\frac{\partial H}{\partial q^i} = -\dot{p}_i. \tag{8.19}$$

To write the last equation, we use the fact that we had previously shown that $\dot{H} = dH/dt = -\partial L/\partial t$. The result is

$$\frac{\partial H}{\partial t} = \dot{H}, \tag{8.20}$$

and we have recovered Eqs. (8.7). These equations are a set of first-order equations in the $\{q^i, p_i\}$ basis. In the next section we explore, in a more geometrical sense, **phase space**, the set on which $\{q^i, p_i\}$ serve as coordinates.

The Hamilton equations (8.18)–(8.20) have a simple form. A simple variational principle can yield the same equations. In fact, the integral $I = \int L \, dt$ still forms the basis of the variational principle. We write

$$I = \int_A^B (p_s \dot{q}^s - H) \, dt \tag{8.21}$$

and consider variations δI where the endpoint times, coordinates, and momenta are held fixed, and variation of momenta is considered separate from and independent from variation of the coordinates. The condition $\delta I = 0$ then gives the Hamilton equations:

$$\delta I = 0 = \int_A^B \left[\left(\dot{q}^s - \frac{\partial H}{\partial p_s} \right) \delta p_s + p_s \delta \dot{q}^s - \frac{\partial H}{\partial q^s} \delta q^s \right] dt. \tag{8.22}$$

The term involving $\delta \dot{q}^s$ is rewritten by integration by parts:

$$\delta I = 0 = \int_A^B \left[\left(\dot{q}^s - \frac{\partial H}{\partial p_s} \right) \delta p_s - \left(\dot{p}_s + \frac{\partial H}{\partial q^s} \right) \delta q^s \right] dt. \tag{8.23}$$

Considering the variations of δp and δq separately then yields Eqs. (8.18) and (8.19); (8.20) is a consequence of the other two.

EXAMPLE

Take

$$L = \tfrac{1}{2} m\, g_{ab}\, \dot{x}^a \dot{x}^b - V(x,t); \tag{8.24}$$

then

$$p_a = m\, g_{ab}\, \dot{x}^b, \tag{8.25}$$

and using the inverse metric we find that

$$H = \frac{1}{2m} g^{ab}\, p_a\, p_b + V(x,t). \tag{8.26}$$

The first term is in fact numerically equal to the kinetic energy, so we see that if the velocities appear only quadratically in the Lagrangian, then the Hamiltonian is equal to the total energy. (Note that in the example just given the total energy is not constant because the Hamiltonian is explicitly time dependent.) In this example it is straightforward to work out the Hamilton equations and see that they are equivalent to the second-order Euler-Lagrange equations.

We have just seen an example of a general result; suppose that $L = T - V$, where T is homogeneous quadratic in the $\dot{q}^i$, and V is independent of $\dot{q}^i$. In this case

$$H = T + V. \tag{8.27}$$

We will prove this result. First, we give an important lemma:

Euler's Theorem on Homogeneous Functions

Let $F(x)$ be a function of several variables $x^1, x^2, \ldots$, which is homogeneous of degree k:

$$F(\lambda x) = \lambda^k F(x). \tag{8.28a}$$

Then

$$x^s \left(\frac{\partial F}{\partial x^s} \right) = k F(x). \tag{8.28b}$$

To prove this result, we differentiate $F(\lambda x)$ with respect to λ:

$$\frac{\partial F(\lambda x)}{\partial \lambda} = \frac{\partial F}{\partial y^i} \frac{\partial y^i}{\partial \lambda}, \qquad where \quad y^i \equiv \lambda x^i$$

$$= \frac{\partial (\lambda^k F)}{\partial \lambda} = k \lambda^{k-1} F(x).$$

Since $\partial y^i / \partial \lambda = x^i$, we have $(\partial F / \partial y^i) x^i = k \lambda^{k-1} F(x)$. Now set $\lambda = 1$ to obtain Euler's theorem.

We now prove that if $L = T - V$, if T is homogeneous quadratic (homogeneous of degree 2) in $\dot{q}^i$, and if V is independent of $\dot{q}^i$, then $H = T + V$. This statement is proved directly from the definition of H and the Euler theorem:

$$H = \dot{q}^i \frac{\partial L}{\partial \dot{q}^i} - L = \dot{q}^i \frac{\partial T}{\partial \dot{q}^i} - T + V = 2T - T + V = T + V.$$

Notice that the definition (8.2) of H in terms of L gives another derivation of $\dot{H} = \partial L / \partial t$: Compute

$$\dot{H} = \dot{q}^i \frac{d}{dt}\left(\frac{\partial L}{\partial \dot{q}^i}\right) + \frac{\partial L}{\partial \dot{q}^i}\ddot{q}^i - \frac{dL}{dt}.$$

However,

$$\frac{dL}{dt} = \frac{\partial L}{\partial \dot{q}^i}\ddot{q}^i + \frac{\partial L}{\partial q^i}\dot{q}^i + \frac{\partial L}{\partial t},$$

so, using the Euler-Lagrange equation, we have

$$\dot{H} = -\frac{\partial L}{\partial t}.$$

8.2 Weiss Action Principle and Noether's Theorem

We assume that there are no nonholonomic constraints. Configuration space $\mathcal{M}$ is an n-dimensional manifold; let $\{q^i\}$ be a coordinate system on it. The Lagrangian is $L(q, \dot{q}, t)$ and is defined for any path $q^i(t)$. Consider the integral **I** defined along this path $\mathcal{P}$:

$$\mathbf{I}[\mathcal{P}] = \int_{t_A}^{t_B} L\, dt, \tag{8.29a}$$

where the path runs from point A at time t_A to point B at time t_B. The value of **I** on a nearby path $\mathcal{P}'$ is

$$\mathbf{I}[\mathcal{P}'] = \int_{t_{A'}}^{t_{B'}} L\, dt, \tag{8.29b}$$

where *we now allow the endpoints (and the end times) to vary*. The integration variable is still labeled t; the path $\mathcal{P}'$ has coordinates $q'^i(t)$. The difference between the values of **I** on these two paths is obtained by writing

$$q'^i(t) \equiv q^i(t) + \varepsilon\eta^i(t), \tag{8.30}$$

$$t'_A \equiv t_A + \varepsilon\tau_A; \qquad t'_B = t_B + \varepsilon\tau_B, \tag{8.31}$$

and expanding $\mathbf{I}[\mathcal{P}'] - \mathbf{I}[\mathcal{P}]$ to first order in ε. We have

$$\mathbf{I}[\mathcal{P}'] - \mathbf{I}[\mathcal{P}] = \int_{t_A + \varepsilon\tau_A}^{t_B + \varepsilon\tau_B} L(q + \varepsilon\eta, \dot{q} + \varepsilon\dot{\eta}, t)\, dt - \int_{t_A}^{t_B} L(q, \dot{q}, t)\, dt \tag{8.32a}$$

$$= \int_{t_A}^{t_B} \left[L(q + \varepsilon\eta, \dot{q} + \varepsilon\dot{\eta}, t) - L(q, q, t) \right] dt$$

$$+ \int_{t_B}^{t_B + \varepsilon\tau_B} L(q, \dot{q}, t)\, dt - \int_{t_A}^{t_A + \varepsilon\tau_A} L(q, \dot{q}, t)\, dt + O(\varepsilon^2). \quad (8.32b)$$

The last two integrals in Eq. (8.32b) are simply (to first order in ε)

$$\varepsilon L\big(q(t_B), \dot{q}(t_B), t_B\big)\tau_B \quad \text{and} \quad \varepsilon L\big(q(t_A), \dot{q}(t_A), t_A\big)\tau_A, \quad (8.33)$$

respectively. We expand the first integral in (8.32b) and integrate by parts:

$$\int_{t_A}^{t_B} \left[L(q + \varepsilon\eta, \dot{q} + \varepsilon\dot{\eta}, t) - L(q, q, t) \right] dt$$

$$= \varepsilon \int_{t_A}^{t_B} \left(\frac{\partial L}{\partial q^i}\eta^i + \frac{\partial L}{\partial \dot{q}^i}\dot{\eta}^i \right) dt \quad (8.34)$$

$$= \varepsilon \int_{t_A}^{t_B} \left(\frac{\partial L}{\partial q^i} - \frac{d}{dt}\frac{\partial L}{\partial \dot{q}^i} \right) \eta^i\, dt + \varepsilon \left(\frac{\partial L}{\partial \dot{q}^i}\eta^i \right)\bigg|_{t_A}^{t_B}.$$

The last expression in (8.34) involves $\partial L/\partial \dot{q}^i$ and η^i evaluated at t_B and at t_A. It is useful, instead of η^i, to use Δq^i, defined as the total difference in the coordinates of the endpoints of the two paths (to first order in ε):

$$\Delta q^i(t_A) = q'^i(t_A') - q^i(t_A)$$

$$= q^i(t_A + \varepsilon\tau_A) + \varepsilon\eta^i(t_A + \varepsilon\tau_A) - q^i(t_A)$$

$$= \varepsilon\dot{q}^i(t_A)\tau_A + \varepsilon\eta^i(t_A), \quad (8.35a)$$

$$\Delta q^i(t_B) = \varepsilon\dot{q}^i(t_B)\tau_B + \varepsilon\eta^i(t_B). \quad (8.35b)$$

The boundary term [the last term in (8.34)] is therefore

$$\varepsilon\left(\frac{\partial L}{\partial \dot{q}^i}\eta^i \right)\bigg|_{t_A}^{t_B} = \left\{ \frac{\partial L}{\partial \dot{q}^i}\big(q(t_B), \dot{q}(t_B), t_B\big) \right\}\left[\Delta q^i(t_B) - \varepsilon\tau_B\dot{q}^i(t_B) \right]$$

$$- \left\{ \frac{\partial L}{\partial \dot{q}^i}\big(q(t_A), \dot{q}(t_A), t_A\big) \right\}\left[\Delta q^i(t_A) - \varepsilon\tau_A\dot{q}^i(t_A) \right]. \quad (8.36)$$

We draw all of these computations together by defining

$$\Delta t(t_A) \equiv \varepsilon\tau_A, \qquad \Delta t(t_B) \equiv \varepsilon\tau_B, \quad (8.37)$$

and by writing

$$\mathbf{I}[\mathcal{P}'] - \mathbf{I}[\mathcal{P}] = \varepsilon \int_{t_A}^{t_B} \left(\frac{\partial L}{\partial q^i} - \frac{d}{dt}\frac{\partial L}{\partial \dot{q}^i} \right)\eta^i\, dt$$

$$+ \left[\left(L - \frac{\partial L}{\partial q^i}\dot{q}^i \right)\Delta t + \left(\frac{\partial L}{\partial \dot{q}^i}\Delta q^i \right) \right]\bigg|_{t_A}^{t_B}. \quad (8.38)$$

The boundary terms are a combination of those from (8.4) and from (8.7).

The **Weiss action principle** is that the system follows the path in configuration space about which general variations produce only endpoint contributions to $\mathbf{I}[\mathcal{P}'] - \mathbf{I}[\mathcal{P}]$. The endpoint contributions that arise are first order and so must be linear combinations of Δq^i (the coefficients turn out to be p_i) and Δt (with coefficient equal to $-H$):

$$\mathbf{I}[\mathcal{P}'] - \mathbf{I}[\mathcal{P}] = (p_i \, \Delta q^i - H \Delta t)\Big|_{t_A}^{t_B}. \tag{8.39}$$

We complete the calculation as in the previous fixed-endpoint case. The integral term in (8.38) must vanish, since it is not an endpoint contribution, implying— because η^i is arbitrary—that the coefficient of η^i in the integrand vanishes:

$$\frac{\partial L}{\partial q^i} - \frac{d}{dt}\left(\frac{\partial L}{\partial \dot{q}^i}\right) = 0, \tag{8.40}$$

and we recover Euler's equations.

Noether's theorem states that if the Lagrangian is invariant under some particular infinitesimal change of coordinates (and time), there are specific associated constants of the motion. We introduce new coordinates and time in a parameterized way, by specifying changes

$$\Delta t = \zeta_s \alpha^s; \qquad \Delta q^i = \zeta_s{}^i \alpha^s,$$

where α^s, $s = 1, \ldots, r$ are the r constant infinitesimal parameters of the transformation, and ζ_s and $\zeta_s{}^i$ are functions of q^i and of t. Thus

$$t' = t + \Delta t, \tag{8.41a}$$
$$q^{i'} = q^i + \Delta q^i \tag{8.41b}$$

are the new coordinates and time parameter.

Suppose that

$$L(t', q^{j'}, \dot{q}^{j'}) \, dt' = L(t, q^j, \dot{q}^j) \, dt. \tag{8.42}$$

Now (8.41a)–(8.41b) are perfectly good choices for variations in time and in the coordinate position, with, thus, $\eta^i = \Delta q^i - \dot{q}^i \Delta t$. Inserting this specific variation gives

$$I[P'] - I[P] = \int_{\varepsilon_A}^{\varepsilon_B} \left(\frac{\partial L}{\partial q^i} - \frac{d}{dt}\frac{\partial L}{\partial \dot{q}^i}\right)\eta^i \, dt + (p_i \, \zeta_s{}^i - H \, \zeta_s)\Big|_{t_A}^{t_B} \alpha^s. \tag{8.43}$$

Because of the invariance stated in (8.42), the left side of this expression vanishes. Assuming then that the path P is an actual solution to the equations of motion, the integral in (8.43) vanishes, and we have a result on certain dynamical quantities; that is,

$$(p_i \, \zeta_s{}^i - H \zeta_s)\alpha^s \tag{8.44}$$

has the same value at time t_B as at time t_A (i.e., it is conserved). Notice that the α^s is an arbitrary parameter and it is the quantities $\zeta_s{}^i$ and ζ_s that define the transformation leaving the Lagrangian invariant.

Notice also that a more general kind of transformation can leave the action invariant, namely, if

$$L(t', q'^i, \dot{q}'^i)\, dt' = L(t, q^i, \dot{q}^i)\, dt + d\left[\phi_s(t, q^j)\alpha^s\right], \qquad (8.45)$$

where the last term is an exact differential and contributes only fixed-endpoint terms in the variational principle. Thus one can state Noether's theorem:

Noether's Theorem

If under the infinitesimal transformation (8.41), the Lagrangian changes by at most an exact infinitesimal differential as in Eq. (8.45), then there are r constants of the motion:

$$\psi_s \equiv H\zeta_s - p_i\, \zeta_s^{\ i} + \phi_s, \qquad s = 1, \ldots, r. \qquad (8.46)$$

Two immediate examples:
(a) Time-independent Lagrangian:

$$t' = t + \alpha, \qquad q'^i = q^i, \qquad \phi = 0.$$

Thus $\zeta_s = 1$, $\zeta_s^{\ i} = 0$;

$$\psi = H = \text{const}.$$

(b) Translation-invariant Lagrangian:

$$t = t', \qquad q'^i = q^i + \alpha^i.$$

Then $\zeta_s = 0$; $\zeta_s^{\ i} = \delta_s^{\ i}$; $\phi = 0$. This gives

$$\psi_s = p_i\, \delta_s^{\ i} = \text{const}.$$

(i.e., conservation of the associated momentum).

8.3   Phase Space and Phase Spacetime

To begin, let us restrict our attention to systems that have no explicit dependence on time:

$$\frac{\partial L}{\partial t} = -\frac{\partial H}{\partial t} = 0. \qquad (8.47)$$

We are accustomed to thinking of a path through configuration space; in coordinates this path is $q^i(t)$. On any such path $L = L\big(q(t), \dot{q}(t)\big)$ and $H = H\big(q(t), p(t)\big)$ are functions of t. L is not simply a function of q^i but depends on the behavior of q^i as a function of t. However, L is a function of the $2n$ variables $q^i, \dot{q}^i$, and once their values are given, L may be computed without the necessity of knowing q^i or $\dot{q}^i$ elsewhere on the path. Similarly, H is best thought of as a function on a $2n$-dimensional manifold with coordinates q^i, p_i.

The coordinates q^i do not typically cover all of configuration space $\mathcal{C}$. We will ignore the question of the global topology of $\mathcal{C}$ in this section, however, and simply consider the patch of configuration space on which a particular coordinate system q^i is valid. When new coordinates q'^i are employed, we will presume that they are defined over the same patch unless the study of the domain of definition raises especially interesting questions. It is important to emphasize those concepts that are independent of any coordinate system. A vector $\mathbf{V}$, for example, may be denoted by its components V^i but is best thought of, as outlined in Chapter 1, as the operator

$$\mathbf{V} = V^i \, \frac{\partial}{\partial q^i},$$

using a notation free of any essential dependence on coordinates.

Phase space is the $2n$-dimensional manifold $\mathcal{P}$ having coordinates q^i, p_j (this is a provisional definition). To emphasize that $\mathcal{P}$ can be thought of as a single entity, we denote its coordinates by z^μ (here $\mu = 1, \ldots, 2n$), where

$$\begin{aligned} z^\mu &= q^\mu & \text{for} \quad \mu &= 1, \ldots, n, \\ z^\mu &= p_{\mu-n} & \text{for} \quad \mu &= n+1, \ldots, 2n. \end{aligned} \tag{8.48}$$

We write $z^\mu = (q^i, p_j)$. At this point we can immediately recognize something strange in Hamilton's equations:

$$\frac{\partial H}{\partial p_i} = \dot{q}^i, \tag{8.49}$$

$$-\frac{\partial H}{\partial q^i} = \dot{p}_i. \tag{8.50}$$

On the right as written here appear the components of a vector in $\mathcal{P}$. In particular, the system follows a path $z^\mu(t)$ through $\mathcal{P}$, namely, the path $q^i(t), p_j(t)$. The tangent vector to this path is

$$\dot{z}^\mu \, \frac{\partial}{\partial z^\mu} = \dot{q}^i \, \frac{\partial}{\partial q^i} + \dot{p}_j \, \frac{\partial}{\partial p_j}, \tag{8.51}$$

and the components $\dot{z}^\mu$ are simply the components of a vector.

On the left of Hamilton's equations, however, appear the components (be careful of the minus sign—see below) of a one-form $\mathbf{d}H$. The components of $\mathbf{d}H$ are

$$H_{,\mu} = \left(\frac{\partial H}{\partial q^i}, \frac{\partial H}{\partial p_j} \right),$$

where $,\mu$ means $\partial/\partial z^\mu$. To make geometric sense out of Hamilton's equations we need a structure that (1) converts $H_{,\mu}$ into the components of a vector like $\dot{z}^\mu$, (2) changes the q components $\partial H/\partial q^i$ into p components so that they can be equated to $\dot{p}_i$, and (3) introduces a minus sign where appropriate.

It is, in fact, a bit simpler to work in the opposite direction: The **symplectic**

2-form is a structure that converts a vector such as $\dot{z}^\mu \partial/\partial z^\mu$ into a one-form, in the process interchanging q and p components and introducing a minus sign. The symplectic form will be defined more precisely below. Its components (in the z^μ coordinate system) are easily defined: for now we define the sympletic form $\mathbf{\Omega}$ as an antisymmetric second-rank tensor having components $\omega_{\mu\nu}$ defined by

$$\omega_{\mu\nu} = \begin{cases} -1 & \text{if } \mu = i, \, \nu = i + n \text{ with } 1 \leq i \leq n, \\ +1 & \text{if } \nu = i, \, \mu = i + n \text{ with } 1 \leq i \leq n, \\ 0 & \text{otherwise.} \end{cases} \tag{8.52}$$

Consequently, $\omega_{\mu\nu}\dot{z}^\nu$ has the following effect: If $\dot{z}^\nu = (0, \dot{p}_j)$, then $\omega_{\mu\nu}\dot{z}^\nu = (-\dot{p}_j, 0)$. If $\dot{z}^\nu = (\dot{q}^i, 0)$, then $\omega_{\mu\nu}\dot{z}^\nu = (0, +\dot{q}^i)$. Hamilton's equations can therefore be written

$$\frac{\partial H}{\partial z^\mu} = \omega_{\mu\nu}\dot{z}^\nu. \tag{8.53}$$

Remember that these equations are written here in a specific coordinate system, $z^\nu = (q^i, p_j)$. Changes of coordinates on $\mathcal{P}$ will be covered later; those that preserve the form of Hamilton's equations by preserving the numerical values of $\omega_{\mu\nu}$ are called **canonical transformations**.

The ways in which the equations of motion are written can be very suggestive. On the one hand, consider Eqs. (8.49) and (8.50): Eq. (8.49) is basically definitional; it is equivalent to

$$p_i = \frac{\partial L}{\partial \dot{q}^i}. \tag{8.54}$$

A path in phase space $q^i(t), p_j(t)$, of course, need not satisfy any particular equation. However, if the path satisfies $\dot{q}^i = \partial H/\partial p_i$, it will be called an **allowed path**. The other Hamilton's equation, Eq. (8.50), is equivalent to the Euler-Lagrange equation

$$\frac{d}{dt}\left(\frac{\partial L}{\partial \dot{q}^i}\right) - \frac{\partial L}{\partial q^i} = 0, \tag{8.55}$$

and in a sense is the embodiment of Newton's second law. An allowed path in phase space that satisfies Eq. (8.55) (i.e., a path that satisfies all of Hamilton's equations) is an **actual path**—namely, a path that the system may actually follow, depending on initial conditions. Written in this manner, the form of Hamilton's equations emphasizes the difference between momentum and coordinate.

When Hamilton's equations are written as Eq. (8.53), however, an even more powerful viewpoint is brought into play. First, the form emphasizes that Hamilton's equations are first order. Consequently, except for pathological cases, a solution is determined by the initial values of q^i and p_j. Put another way: Through each point in phase space, $\mathcal{P}$, passes just one actual path.

Second, it is $\omega_{\mu\nu}$ that distinguishes "momentum" from "coordinate." We will see later that a canonical transformation, one preserving the form of $\omega_{\mu\nu}$, may prove to have powerful usage. Such a transformation will, perhaps, mix up our preconceived notions of momentum and coordinate, yet preserve a more basic distinction as embodied in the symplectic form.

Finally, the foregoing form shows that the phase space path velocity is related to a gradient. Suppose, for example, we were given the equations of motion in the form

$$\dot{z}^{\mu} = F^{\mu}(q, p), \tag{8.56}$$

where F^{μ} is simply a vector. To check whether these equations are of Hamilton's type, we compute

$$(\omega_{\mu\nu} F^{\mu})_{,\lambda} - (\omega_{\mu\lambda} F^{\mu})_{,\nu},$$

where $,\mu$ means $\partial/\partial z^{\mu}$. This antisymmetric derivative vanishes when $\omega_{\mu\nu} F^{\nu} = H_{,\mu}$. Moreover, if this derivative vanishes, then we know, in a limited region, at least, that a function H does exist such that $\omega_{\mu\nu} F^{\nu} = H_{,\mu}$. This subject is treated in more detail in Chapter 10.

When the Hamiltonian has explicit time dependence, most of the remarks above follow directly. Nevertheless, it pays to reconsider the structure of phase space from the spacetime point of view. Configuration spacetime $\mathcal{C}^*$ is the set defined by coordinates (q^i, t), at least as a provisional definition. Let τ be the path parameter, so that $\big(q^i(\tau), t(\tau)\big)$ defines a path in spacetime. We normally take $t(\tau) = \tau$, but the notation here is meant to distinguish between t as time used as a coordinate in spacetime and τ as time used as a path parameter. The same path extremizes the action, no matter what the parameter.

Now, of course, $L(\dot{q}, q, t)$ remains the Lagrangian when used as in integral of the form

$$I = \int_A^B L\Big(\frac{q'}{t'}, q, t\Big) t' \, d\tau, \tag{8.57}$$

with prime denoting $d/d\tau$. The Euler-Lagrange equations are given in terms of the integrand $\tilde{L} = Lt'$:

$$\frac{d\tilde{p}_i}{d\tau} - \frac{\partial \tilde{L}}{\partial q^i} = 0, \tag{8.58}$$

$$\frac{d\tilde{p}_t}{d\tau} - \frac{\partial \tilde{L}}{\partial t} = 0, \tag{8.59}$$

where we have defined $\tilde{p}_i$ and $\tilde{p}_t$ to be, respectively, the momenta conjugate to q^i and to t:

$$\tilde{p}_i = \frac{\partial \tilde{L}}{\partial q'^i}, \qquad \text{where} \quad q'^i = \frac{dq^i}{d\tau}, \tag{8.60}$$

$$\tilde{p}_t = \frac{\partial \tilde{L}}{\partial t'}, \qquad \text{where} \quad t' = \frac{dt}{d\tau}. \tag{8.61}$$

The explicit form of $\tilde{p}_i$ is derived using the chain rule (treating $\dot{q}^i$ as the function q'^i/t' and treating $\tilde{L} = Lt'$ with L as a function of $\dot{q}^i$):

$$\tilde{p}_i = \frac{\partial \tilde{L}}{\partial q'^i} = t' \frac{\partial L}{\partial \dot{q}^s} \frac{\partial \dot{q}^s}{\partial q'^i} = \frac{\partial L}{\partial \dot{q}^i} = p_i. \tag{8.62}$$

Similarly, we find that

$$\tilde{p}_t = \frac{\partial \tilde{L}}{\partial t'} = L + t' \frac{\partial L}{\partial t'} = L + t' \frac{\partial L}{\partial \dot{q}^s} \frac{\partial \dot{q}^s}{\partial t'}$$

$$= L - \frac{q'^s p_s}{t'} = L - \dot{q}^s p_s. \tag{8.63}$$

The Euler-Lagrange equations then can be rewritten as

$$\frac{dp_i}{d\tau} - t' \frac{\partial L}{\partial q^i} = 0, \tag{8.64}$$

$$\frac{dp_t}{d\tau} - t' \frac{\partial L}{\partial t} = 0, \tag{8.65}$$

where we now set $p_t = \tilde{p}_t$. When we recognize

$$p_t = -H \tag{8.66}$$

and

$$\frac{d}{d\tau} = t' \frac{d}{dt}, \tag{8.67}$$

we see that they are the same as the usual Euler-Lagrange equations.

We next attempt to develop the Hamiltonian framework. However, we find a difficulty. If we define $\tilde{H}$ in terms of $\tilde{L}$ by

$$\tilde{H} = \tilde{p}_i q'^i + \tilde{p}_t t' - \tilde{L}, \tag{8.68}$$

we find that

$$\tilde{H} = p_i \dot{q}^i t' - Ht' - Lt' = 0. \tag{8.69}$$

Using this straightforward method, therefore, we have found a Hamiltonian that vanishes identically, and the Hamilton equations clearly are not correct.

The reason for $\tilde{H} = 0$ is that the integral $I = \int \tilde{L} \, d\tau$ is independent of the parameter τ in the sense that an arbitrary change of parameter to any other makes no difference in the value of I. A similar difficulty would happen in a much simpler situation if we chose $\sqrt{T}$ instead of T (kinetic energy) for the Lagrangian of a free particle. With T as the Lagrangian, the equations of motion show that the particle follows a straight line at constant speed. With $\sqrt{T}$ as the Lagrangian, the orbit (a straight line) is found, but since in this case $\int L \, dt$ is independent of path parameter, no requirement on the speed can be found. The Hamiltonian that would be derived from the Lagrangian $\sqrt{T}$ is identically zero, just as is $\tilde{H}$.

There is a way around this difficulty. Equation (8.66) shows that

$$0 = p_t + H(q_1, \ldots, q^{n+1} ; \; p_1, \ldots, p_n), \tag{8.70}$$

where $q^{n+1} = t$. Equations (8.60) and (8.61) can obviously not be solved for the q'^i, since rescaling (for instance) the parameter τ rescales the q'^i, but the variational principle is by construction invariant under changes in parameterization. Now if Eqs. (8.60) and (8.61) are not solvable, it must mean that they are not functionally independent of each other. There must therefore be a functional relation between the coordinates q^α (these include q^i and t) and the momenta p_α (including the p_i and p_t). Equation (8.70) is precisely this relation.

The difficulty with the correctness of the Hamilton equations, namely, that in the variational principle of the type of

$$0 = \delta \int \sum_{\alpha=1}^{n+1} p_\alpha \, q'^\alpha \, d\tau,$$

no Hamiltonian appears, is corrected by noting that the constraint (8.70) must be fulfilled. By introducing a Lagrange multiplier, the correct variational principle becomes

$$0 = \delta \int \sum_{\alpha=1}^{n+1} (p_\alpha \, q'^\alpha - \lambda K) \, d\tau, \tag{8.71}$$

where $K = 0$ expresses the content of Eq. (8.70). [After a canonical transformation, (8.70) may not have such a simple form. Hence the more elaborate notation.] If the constraint is explicitly eliminated, meaning we consider only motions and variations that satisfy $K = 0$, this last term may be dropped from the action, and we explicitly eliminate $p_{n+1} = p_t$ in favor of its expression (8.70) in terms of other dynamical variables:

$$\begin{aligned} 0 &= \delta \int (p_i \, q'^i - Ht') \, d\tau \\ &= \delta \int (p_i \, \dot{q}^i - H) \, dt. \end{aligned} \tag{8.72}$$

On the other hand, we may note that τ in Eq. (8.71) is a completely arbitrary variable, and we may, for every path, make a parameter choice determined by the requirement $\lambda = 1$. Then the variational principle is

$$0 = \delta \int \sum_{\alpha=1}^{n+1} (p_\alpha \, q'^\alpha - K) \, d\tau, \tag{8.73}$$

and K [called the **extended Hamiltonian** by Lanczos (*Variational Principles of Mechanics,* 4th ed., U. of Toronto Press, Toronto, 1970)] plays the role of a Hamiltonian in this extended system. Now the parameter τ, being totally arbitrary, appears explicitly nowhere in the variational principle. Therefore,

$$\frac{\partial K}{\partial \tau} = \frac{dK}{d\tau} = 0. \tag{8.74}$$

Hence K persists at any constant initial value; the value $K = 0$ implied by the constraint (8.70) is maintained, but the functional form of K allows meaningful equations of motion to be derived, as may be verified by direct calculation. The Hamilton equations are

$$q'^{\alpha} = \frac{\partial K}{\partial p_{\alpha}}, \qquad p'_{\alpha} = -\frac{\partial K}{\partial q^{\alpha}}. \tag{8.75}$$

For the special choice $K = p_{n+1} + H$,

$$q'^{i} = \frac{\partial H}{\partial p_i}, \tag{8.76a}$$

$$p'_i = -\frac{\partial H}{\partial q_i}, \tag{8.76b}$$

$$q'^{n+1} = 1, \tag{8.76c}$$

$$p'_{n+1} = -\frac{\partial H}{\partial q^{n+1}}. \tag{8.76d}$$

Equation (8.76c) shows that the parameterization is recovered in terms of t; for the particular choice of K, τ has again become explicitly t, and p_{n+1} has become $-H$. The final Eq. (8.76d) states once again that the time dependence of the Hamiltonian H arises only from the explicit appearance of t.

We may elaborate on the difference between the vanishing Hamiltonian that arose in Eq. (8.69) and the constraint (8.70). The important point is that (8.69) is an equation that contains p, q, and q', so it is not legitimately in the form of a Hamiltonian. But its value is identically zero, no matter what motion we might imagine, so there is no way to reduce it in such a way as to eliminate q' (without in fact setting it to zero); and anyway we have seen that q' cannot be solved for from the Euler-Lagrange equations involving $\tilde{L}$. On the other hand, Eq. (8.70) is explicitly a function only of the p_i and the q^i. Even though the actual conserved value of K is zero, one can just as well imagine solving the equations of motion for any other value of K. The constraint that K vanish is of entirely different character than the result (8.69) that $\tilde{H}$ vanishes.

This extended Hamiltonian formulation of mechanics has important implications for the study of relativistic systems, say, or any parameter-independent formulation of mechanics. Notice that in particular it says that every Hamiltonian, conservative or not, can be replaced by a conservative Hamiltonian system with a phase space of two higher dimensions. The physical motion is required to lie in a $(2n + 1)$-dimensional subspace of the $(2n + 2)$-dimensional phase spacetime, but this requirement is not a constraint. It is, instead, simply a restriction on the initial value of the extended Hamiltonian. It is a restriction similar to studying simple harmonic motion with one given energy. It introduces absolutely no mathematical or physical complications.

Finally, we mention a procedure for defining a type of Hamiltonian involving only some variables. Suppose that we define H_k by

$$H_k \equiv \sum_{i=1}^{k} p_i \, \dot{q}^i - L \qquad \left(\text{where } p_i = \frac{\partial L}{\partial \dot{q}^i}; \quad i = 1, \ldots, k \right).$$

We shall not prove it here, but the Euler-Lagrange equations may be expressed as

$$\frac{\partial H_k}{\partial p_i} = \dot{q}^i, \quad \frac{\partial H_k}{\partial q^i} = -\dot{p}_i, \qquad \text{for} \quad i = 1, \ldots, k,$$

$$\frac{d}{dt} \left(\frac{\partial H_k}{\partial \dot{q}^a} \right) - \frac{\partial H_k}{\partial q^a} = 0, \qquad \text{for} \quad a = k+1, \ldots, n.$$

H_k is to be expressed as a function of p_i, q^i (for $i = 1, \ldots, k$) and of $\dot{q}^a$, q^a (for $a = k+1, \ldots, n$), and possibly of t. Instead of spacetime $\mathcal{C}^*$ [$(n+1)$-dimensional], we think of H_k as pertaining to a manifold with coordinates p_i, q^i ($i = 1, \ldots, k$), q^a ($a = k+1, \ldots, n$), and t [i.e., to a manifold with $2k + (n + 1 - k) = n + 1 + k$ dimensions]. The equations of motion on this manifold are partly first order and partly second order.

To handle the first-order equations, we define

$$z^\mu \equiv q^\mu \qquad \text{for} \quad \mu = 1, \ldots, k,$$
$$z^\mu \equiv p_{\mu-k} \qquad \text{for} \quad \mu = k+1, \ldots, 2k.$$

The symplectic form $\mathbf{\Omega}$ with components $\omega_{\mu\nu}$ is defined as before, but it involves only indices $\mu, \nu = 1, \ldots, 2k$. Coordinates z^μ for $\mu = 2k + 1, \ldots, n + 1 + k$ are defined to be the q^a ($a = k+1, \ldots, n$) and t, but $\omega_{\mu\nu}$ has only zero values if μ or ν goes from $2k = 1$ to $n + 1 + k$. It is clear that this procedure suffers from a bit of complexity.

8.4   Examples

A simple time-independent system whose phase space $\mathcal{P}$ is easy to analyze is the harmonic oscillator. The Hamiltonian is

$$H = \frac{1}{2m} p^2 + \frac{m}{2} \omega^2 x^2, \tag{8.77}$$

and the resultant equations of motion are

$$\dot{x} = \frac{p}{m}, \qquad \dot{p} = -m\omega^2 x. \tag{8.78}$$

$\mathcal{P}$ is two-dimensional, and much information can be obtained by plotting $(\dot{z}^1, \dot{z}^2)$ $= (p/m, -\omega^2 m x)$. In particular, $z^\alpha = 0$ at $p = x = 0$; z^1 vanishes at $p = 0$, and z^2 vanishes at $x = 0$. The integral curves in z^α describe circles about the origin, and as time evolves the system periodically traces out one of these circles. The motion of the system can be read off a diagram of the phase space. This qualitative investigation of $\mathcal{P}$ often yields results when no analytical solution of the system is available. Because the Hamiltonian has no explicit time dependence, the Hamiltonian is itself a constant of the motion (equal to the energy).

The importance of constants of the motion cannot be overemphasized for the solution of mechanical problems. In the harmonic oscillator example we could have restricted our consideration to that circle on which the energy took the value given by our initial data. The existence of constants has the effect of reducing the dimensionality of the available phase space by one for each constant. In a one-dimensional time-independent mechanical system (with two-dimensional phase space), the determination of two constants of the motion localizes the system point to a zero-dimensional subset of the phase space (i.e., to a point or a collection of points). It is of interest to see how this comes about in the harmonic oscillator problem. Clearly, the energy provides a constant, which restricts us to the one-dimensional subset that is the circle $E = $ constant. But what of the other constant of the motion? To identify it, let us obtain the complete solution to the mechanical problem.

We use the constancy of the Hamiltonian, which implies that

$$\frac{p^2}{2m} = E - \frac{1}{2}m\omega^2 x^2. \tag{8.79}$$

Since $p = m\dot{x}$, we have

$$\left(\frac{2}{m}E - \omega^2 x^2\right)^{-1/2} dx = dt. \tag{8.80}$$

Integration yields

$$-\frac{1}{\omega}\arccos\left[\left(\frac{m\omega^2}{2E}\right)^{1/2} x\right] = (t - t_0). \tag{8.81}$$

Taking the cosine of each side gives

$$x = \left(\frac{2E}{m\omega^2}\right)^{1/2}\cos\omega(t - t_0). \tag{8.82}$$

The final form of the solution depends on the constant energy E and on another constant, t_0. Notice that t_0, the constant of the motion, is given by

$$t_0 = t - \frac{1}{\omega}\arccos\left[\left(\frac{m\omega^2}{2E}\right)^{1/2} x\right], \tag{8.83}$$

which is the expression of the constant t_0 in terms of x and p (via E) and in terms of t. A choice of t_0 thus gives a definite relation between t and x, which localizes the system point in the phase space. It is clear that the two constants of the motion are qualitatively different; the energy E has no explicit time dependence, whereas t_0 does; t_0 could be obtained only by explicit integration of the equation of motion.

The simple pendulum provides a second example of a two-dimensional phase space. The position, θ, of the pendulum describes its position on a circle defined by the length ℓ of the (rigid) rod supporting the mass. By constructing the Lagrangian $L = T - V$,

$$L = \tfrac{1}{2}m\ell^2\dot{\theta}^2 - mg\ell(1 - \cos\theta), \tag{8.84}$$

we find that

$$H = p_\theta\dot{\theta} - L = (2m\ell^2)^{-1}p_\theta{}^2 + mg\ell(1 - \cos\theta), \tag{8.85}$$

where

$$p_\theta = m\ell^2\dot{\theta} \tag{8.86}$$

and g is the acceleration of gravity. Because of this definition of the momentum, p_θ can apparently take any value, $p_\theta \in (-\infty, +\infty)$. On the other hand, we have already mentioned that the configuration space has the topology of the circle $\mathcal{S}^1$. The topology of the phase space for the pendulum is thus cylindrical:

$$\mathcal{P} = \mathcal{S}^1 \times \mathbf{R}. \tag{8.87}$$

Figure 8.1 is a diagram of the cylinder, but it is often more convenient to roll out the cylinder onto the plane. The cylinder may be cut at any value of θ_{cut} to accomplish this (see Figure 8.2). To eliminate the apparent discontinuities that occur when the pendulum moves through θ_{cut}, one may repeat the copy of the cylinder onto the plane (Figure 8.3). Since the Hamiltonian is time independent, the orbits in phase space will be lines of constant H.

It is well known that for $\theta \ll 1$ the pendulum reduces to the harmonic oscillator problem. Hence if $|\theta_{\text{max}}| \ll 1$, we see the ellipses of the harmonic oscillator phase space, centered on the origin in $\mathcal{P}$. Figure 8.3 gives the quantities $\dot{\theta}, \dot{p}$ as vectors in phase space.

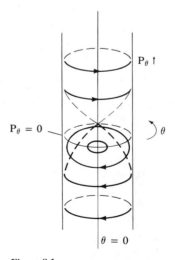

$P_\theta \uparrow$

$P_\theta = 0$

θ

$\theta = 0$

Figure 8.1

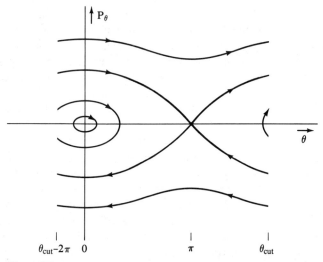

Figure 8.2

A moment's reflection shows that the maximum restoring torque on the pendulum occurs for $\theta = \pm\pi/2$, and the restoring force is always less than that obtained in the small-angle harmonic oscillator approximation. Hence when θ is not small, the momentum decreases less rapidly than for the harmonic oscillator situation. The result is that the closed $H = $ constant curves are elongated in the θ-direction.

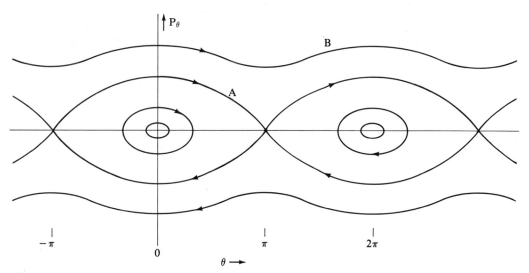

Figure 8.3

Of particular interest is the curve marked A in Figure 8.3, which describes a motion in which the pendulum has $p_\theta = 0$ (i.e., vanishing velocity) when $\theta = \pi$, the top of the pendulum. This point ($\theta = \pi, p_\theta = 0$) is clearly a point of unstable equilibrium, and the pendulum may leave this point by one of the two paths (indicated by the direction of the arrows) leading away from it.

Note that since it is an unstable equilibrium, we have $p_\theta \propto e^{\alpha t}$ for motion away from that point and $p_\theta \propto e^{-\alpha t}$ for motion toward it, but then since $d(\pi - \theta)dt \propto p_\theta$, we have $(\pi - \theta) \propto \pm p_\theta$. This is the reason the curves A cross at ($\theta = \pi, p_\theta = 0$) with nonzero slope and nonzero angle between them. [Alternatively, calculate the constant value of H to be $2mg\ell$ from the requirement $p_\theta = 0$ when $\theta = \pi$ and express H in the region $\theta \approx \pi$. The result will be that $p_\theta{}^2 \propto (\theta - \pi)^2$.]

If we supply the pendulum with more than $2mg\ell$ total energy, its kinetic energy will never vanish and it will continue forever with the same sign of $\dot\theta$. These are the curves marked B and C, and they become more nearly straight as the energy increases; the gravitational potential becomes a mere perturbation.

An allowed path in $\mathcal{P}$ is one that obeys the definitional law $p = \partial L/\partial\dot\theta$. This law says $m\ell^2\dot\theta = p_\theta$. Consequently, any path of the form $\theta(t), p_\theta(t)$ such that $m\ell^2\dot\theta = p_\theta$ is an allowed path. Only when the path obeys $\dot p = -\partial H/\partial\theta$ is the path an actual one. Consider, for example, the path $\theta = \cos t, p = At$, where $A = $ const. Clearly, no choice of A makes this path either allowed or actual. On the other hand, consider the path $\theta = A\cos t, p = B\sin t$. For certain values of A and B this is an allowed path. However, it is clearly not an actual path; it is a physically allowed path but not one conforming to the specific forces involved. This allowed-but-not-actual path has the pendulum moving like a simple harmonic oscillator; the former neither-actual-nor-allowed path cannot be pictured as a physical motion of the pendulum.

We now turn to the example of finding a geodesic in space. Such a curve is the shortest line joining two points, and of course it is a straight line. The variational principle is to minimize the integral of ds (s being arc length) between two points A, B:

$$\delta \int_A^B ds = 0. \tag{8.88}$$

We let λ be an arbitrary path parameter, so that

$$ds = \left[\left(\frac{dx^1}{d\lambda}\right)^2 + \left(\frac{dx^2}{d\lambda}\right)^2 + \left(\frac{dx^2}{d\lambda}\right)^2\right]^{1/2} d\lambda. \tag{8.89}$$

The integrand is thus

$$L = \left[\left(\frac{dx^1}{d\lambda}\right)^2 + \left(\frac{dx^2}{d\lambda}\right)^2 + \left(\frac{dx^2}{d\lambda}\right)^2\right]^{1/2}, \tag{8.90}$$

and the Euler-Lagrange equations are (with $\alpha = 1, 2, 3$)

$$\frac{dp_\alpha}{d\lambda} = \frac{d}{d\lambda}\left(\frac{1}{L}\frac{dx^\alpha}{dt}\right) = 0. \tag{8.91}$$

The solution is clearly

$$\frac{1}{L}\frac{dx^\alpha}{d\lambda} = c^\alpha = \text{constant such that } (c^1)^2 + (c^2)^2 + (c^3)^2 = 1. \qquad (8.92)$$

This path is a straight line, since $dx^\alpha/d\lambda \propto dx^\beta/d\lambda$, but notice that the path parameter is arbitrary.

The arbitrariness in λ implies that the Hamiltonian derived from L is zero. It also shows up when we compute (˙ means $d/d\lambda$):

$$\frac{\partial^2 L}{\partial \dot{x}^\alpha \partial \dot{x}^\beta} = \left(\frac{1}{L}\right)^3 (L^2 \delta_{\alpha\beta} - \dot{x}^\alpha \dot{x}^\beta). \qquad (8.93)$$

It is easy to compute the determinant of this 3×3 matrix and to see that it is zero (see Exercise 8.14). Consequently, it is not possible to replace $\dot{x}^\alpha$ by the conjugate "momenta" $L^{-1}dx^\alpha/d\lambda$ in a Hamiltonian formalism. Notice that (8.92) contains a condition [required so that the definition of L in (8.91) is maintained]. This condition is analogous to the condition (8.70) in the spacetime formulation of Hamiltonian mechanics. We have

$$K = 1 - \sum_{\alpha=1}^{3} p_\alpha \, p_\alpha = 1 - \sum_{\alpha=0}^{3} c^\alpha c^\alpha = 0.$$

The variational principle—written in a Hamiltonian form—thus is

$$\delta \int (p_\alpha \dot{x}^\alpha - \Lambda K)\, d\lambda = 0,$$

where we are now using Λ as the multiplier. If we rescale the parameter so that $d\tilde{\lambda} = \Lambda\, d\lambda$ and drop the ˜, Λ is eliminated, and the variational principle yields the Hamilton's equations appropriate to K as a Hamiltonian:

$$\delta \int (p_\alpha \dot{x}^\alpha - K)\, d\lambda = 0;$$

that is,

$$\frac{dp_\alpha}{d\lambda} = -\frac{\partial K}{\partial x^\alpha} = 0,$$

$$\frac{dx^\alpha}{d\lambda} = \frac{\partial K}{\partial p_\alpha} = p_\alpha.$$

The motion is not completely specified only by these equations. Notice that $\partial K/\partial \lambda = 0$; hence K maintains its initial value, which physically is required to be zero. For this problem, notice finally that L^2 is the Lagrangian for an ordinary free particle (of mass $= 2$), and no difficulty at all occurs with the Hamiltonian if we use L^2 as the Lagrangian.

As an example involving explicit time dependence, consider the case of a particle that slides without friction on a straight wire, which rotates about an axis perpendicular to itself. Let the wire rotate in the $(x^1$–$x^2)$-plane, so that at time t the position of the mass m is

$$x^1 = q(t)\cos\theta(t),$$
$$x^2 = q(t)\sin\theta(t),\tag{8.94}$$

where $\theta(t)$ is a given function of t and q is the single coordinate needed to describe configuration space. The Lagrangian L is

$$L = T = \frac{m}{2}[(\dot{x}^1)^2 + (\dot{x}^2)^2] = \frac{m}{2}(\dot{q}^2 + q^2\dot{\theta}^2).\tag{8.95}$$

The momentum p is given by

$$p = m\dot{q},\tag{8.96}$$

so that

$$H = \frac{1}{2m}p^2 - \frac{m}{2}q^2\dot{\theta}^2.\tag{8.97}$$

Notice that H is not the same as total energy. Notice that it is possible for H to be a constant of the motion, namely, when $\dot{\theta} = \text{const.}$:

$$\dot{H} = \frac{\partial H}{\partial t} = -mq^2\dot{\theta}\ddot{\theta}.\tag{8.98}$$

[For example, if $\theta = \omega t$ (where $\omega = \text{const.}$), $\dot{H} = 0$.] The Hamilton equations are (8.98) and

$$\dot{q} = \frac{\partial H}{\partial p} = \frac{p}{m},\tag{8.99}$$

$$\dot{p} = -\frac{\partial H}{\partial q} = mq\dot{\theta}^2.\tag{8.100}$$

To take a specific case, let $\theta = \sqrt{2}\,\log t$, so that the rotating wire slows its rotation as t increases, asymptotically halting but always turning. Then $\dot{\theta} = \sqrt{2}/t$. The general solution for $q(t), p(t)$ is

$$q(t) = At^2 + \frac{B}{t}, \qquad p(t) = \left(2At - \frac{B}{t^2}\right)m \qquad (A, B \text{ const.}).$$

Then, along any of this two-parameter set of actual paths, $H(t)$ has the value

$$H = mA^2t^2 - \frac{B^2m}{2t^4}.$$

When the Lagrangian depends on the velocities in a way that is not just quadratic, the Hamiltonian can still be computed if none of the velocities appear only linearly in the Lagrangian. If the Lagrangian does depend on a velocity $\dot{q}$ only linearly, then $p = \partial L/\partial \dot{q}$ gives a right-hand side that is independent of $\dot{q}$ and thus yields no relation defining p in terms of $\dot{q}$. The Legendre transformation can be worked in the opposite direction, using $\dot{q} = \partial H/\partial p$, but a putative Hamiltonian that is linear in the momenta cannot be directly related to a Lagrangian. This "exceptional" case is in fact of great importance and is discussed briefly in Chapter 10.

An important example in which the Lagrangian has terms linear in the velocity as well as other terms is that for a particle in an **electromagnetic field**. We have already written the Lagrangian for that system [Eq. (7.11)]. We obtain

$$p_i = m(\delta_{is}\dot{x}^s + eA_i), \tag{8.101}$$

which implies that

$$\dot{x}^k = \frac{1}{m}\delta^{ks}(p_s - eA_s). \tag{8.102}$$

Hence

$$\begin{aligned} p_i\dot{x}^i - L &= (m\delta_{is}\dot{x}^s + eA_i)\dot{x}^i - \tfrac{1}{2}\delta_{is}\,m\dot{x}^i\dot{x}^s - eA_s\,\dot{x}^s + e\phi \\ &= \tfrac{1}{2}\delta_{is}\,\dot{x}^i\dot{x}^s + e\phi. \end{aligned} \tag{8.103}$$

When written correctly in terms of p_i instead of $\dot{x}^i$, we find that

$$H = \frac{1}{2m}\delta^{is}(p_i - eA_i)(p_s - eA_s) + e\phi. \tag{8.104}$$

Notice that the presence of a term linear in the velocity does not prevent inversion of the defining equation for p_i. Also notice that the resultant Hamiltonian is numerically equal to the sum of the kinetic energy and electrostatic potential energy of the particle. Because the magnetic interaction can do no work on the particle, this sum could be called the energy of the particle; it will be conserved only if A_i and ϕ are time independent. Note that if we claimed $L = T - V$ (with a velocity-dependent potential), then $H \neq T + V$.

Finally, note the structure $p_i - eA_i$ in the Hamiltonian. The assumption that electromagnetism couples to the Hamiltonian so that the vector potential appears only in this combination with the momentum is called the "minimal coupling" prescription. The Hamilton equations are

$$\dot{p}_i = -\frac{\partial H}{\partial x_i} = \frac{1}{m}A_{s,i}\,\delta^{st}(p_t - eA_t) - e\phi_{,i}, \tag{8.105}$$

and

$$\dot{x}^i = \frac{\partial H}{\partial p_i} = \frac{1}{m}\delta^{is}(p_s - eA_s). \tag{8.106}$$

This second equation is an identity [see Eq. (8.102)]. If the expression for $\dot{x}^i$ is solved for p_i and inserted into the p_i equations, we obtain

$$m\ddot{x}^s\delta_{si} + e\frac{dA_i}{dt} = eA_{s,i}\,\dot{x}^s - e\phi_{,i}. \tag{8.107}$$

Now A_i has a time derivative because it may depend explicitly on time and because it depends on x^i. Hence

$$\begin{aligned} m\ddot{x}^s\delta_{si} &= -e\left(\phi_{,i} + \frac{\partial A_i}{\partial t}\right) + e(A_{s,i} - A_{i,s})\dot{x}^s \\ &= e(\mathbf{E})_i + e(\mathbf{v}\times\mathbf{B})_i. \end{aligned} \tag{8.108}$$

This equation shows how the **Lorentz force** depends on the electric field $\mathbf{E}$ and the magnetic field $\mathbf{B}$.

EXERCISES

8.1. In developing the Hamilton equations we assumed that $\partial^2 L/\partial \dot{q}^i \, \partial \dot{q}^j$ is nonsingular. Suppose we do not make this assumption; what (weakened) form of the Hamilton's equations result if
(a) $\partial^2 L/\partial \dot{q}^i \, \partial \dot{q}^j$ commutes with $\partial^2 L/\partial \dot{q}^i \, \partial q^j$?
(b) $\partial^2 L/\partial \dot{q}^i \, \partial q^j$ is noninvertible?
(c) $\partial^2 L/\partial \dot{q}^i \, \partial q^j = 0$?
Assumptions will also have to be made about $\partial^2 L/\partial \dot{q}^i \, \partial t$. In part (c), give an interpretation of the equations of motion so obtained.

8.2. As an example of part (c) of Exercise 8.1, consider the Lagrangian

$$L = \dot{x}^2 + (\dot{x} - \dot{y})B,$$

with B a given constant. Show that in this case, the equations give constant p_x, p_y, even though $\dot{y}$ cannot be defined in terms of p_y.

8.3. Suppose that we have a Hamiltonian system, obtained so that Eqs. (8.18)–(8.20) are satisfied. Then one could view the equation

$$L = p\dot{q} - H$$

as defining a Lagrangian for the system; and the Hamilton equations would imply the Lagrange equations, at least if certain second-derivative matrices were nonsingular. If the fundamental variables are assumed to be r, p, τ, and we have the relations

$$q^i = r^i, \qquad \dot{q}^i = \frac{\partial H}{\partial p_i}, \qquad t = \tau,$$

together with Eq. (8.9), obtain the Lagrange equations and specify what conditions are necessary on H for them to hold.

8.4. A particle in a uniform gravitational field is constrained to the surface of a sphere centered at the origin. The radius of the sphere varies: $r = r(t)$ with $r(t)$ a given function. Obtain the Hamiltonian and the canonical equations. Discuss energy conservation. Is the Hamiltonian the total energy?

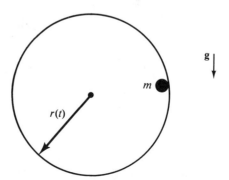

Figure P8.4

8.5. Consider force-free motion of a mass m in a plane. Introduce new coordinates q^1, q^2 related to the rectangular x, y by $x^2 - y^2 = q^1$, $x + y = q^2$. Show that this coordinate transformation is nonsingular except on the line $x = -y$. Express the kinetic energy in terms of the coordinates q^1, q^2. Obtain the Lagrangian, the Hamiltonian, and the equations describing force-free motion, in terms of the coordinates q^1, q^2. Solve the equations in this coordinate system.

8.6. Suppose that the Lagrangian of a system is independent of one particular coordinate, say, x^1, and also is time independent.
  (a) Transform to moving coordinates $\tilde{x}^1 = x^1 - at$. Is the Hamiltonian in the new coordinates still conserved? How does the new Hamiltonian compare to the old?
  (b) As a concrete example of the situation in part (a), consider the case of an axisymmetric system of one particle (V independent of angle θ, using cylindrical coordinates). Transform to rotating coordinates $\tilde{\theta} = \theta - \omega t$. Describe the new Hamiltonian and show that it is not the total energy, but rather, differs from the total energy by a significant term.
  (c) Knowing the answer to part (b), do part (b) again in Cartesian coordinates x, y, z.

8.7. Describe the motion of a particle with a Hamiltonian

$$H(p, x) = \tfrac{1}{2}p^2 + \tfrac{1}{2}\omega^2 x^2 + \lambda(\tfrac{1}{2}p^2 + \tfrac{1}{2}\omega^2 x^2)^2,$$

where ω and λ are constants.

8.8. The Hamiltonian for a free particle moving in one dimension is $H = (1/2m)\,p^2$. Use the Hamilton equation in terms of the symplectic form, Eq. (8.77), to obtain a picture of the motion in phase space. Sketch a diagram, showing the lines of constant H and the integral lines of the vector $\dot{z}$ obtained in this way. What changes if the motion is confined to $x > 0$ by a reflecting wall?

8.9. Consider a particle of mass m in a one-dimensional potential of the form $k\left[\exp(x^2) - 1\right]$.
  (a) Plot the potential (sketch only) and qualitatively describe motion in this potential for both small and large amplitudes. In particular, find the frequency of motion for small amplitudes $x^2 \ll 1$ and also for the opposite case $x^2 \gg 1$. For the small-amplitude case, at least, give an exact result. For the large-amplitude case it is also straightforward to obtain the period correct to order $x^2\left[1 + \mathcal{O}(1/x)\right]$. At any rate, give a discussion for the results expected in the large-amplitude case.
  (b) Derive the Lagrangian and the Hamiltonian for the system defined above. Use the Hamiltonian equations to plot a flow in the phase space of this system and describe the motion of the system in phase space. Take some care in drawing these figures; numerical calculations are not needed, but the figure should accurately reflect the qualitative features of the phase space.

8.10. A charged particle moves in a magnetic field. Its Hamiltonian is

$$H = \frac{1}{2m}\delta^{st}(p_s - eA_s)\,(p_t - eA_t).$$

Obtain the Lagrangian. Set up Hamilton's equations. Show that gauge transformations change the Hamiltonian and the orbits in phase space, but not the orbits in configuration space. Now suppose that $A_i = K\delta_{is}\,x^s$, K = constant. Find the motion (i.e., solve the Hamilton equations).

8.11. Consider the Lagrangian

$$L = \tfrac{1}{2}me^{\lambda t}(\dot{x}^2 - \omega^2 x^2),$$

where m, λ, and ω are constants. Write the variational principle and find the equation of motion. What does it correspond to? Find the Hamiltonian. Is it conserved? Rewrite the variational principle in a way that includes t as a coordinate. What is the momentum conjugate to t? What is the variational principle written in terms of the extended Hamiltonian K? (Give the explicit form of K and give the Hamilton equations in this form.)

8.12. Consider the Lagrangian

$$L = \tfrac{1}{2}\dot{x}^2 - \tfrac{1}{2}\omega_0^2(1 + at)^2 x^2.$$

Obtain the Hamiltonian and the Hamilton equations for the Lagrangian above via a variational principle. Is the Hamiltonian conserved? Rewrite the variational principle in a way that includes t as a coordinate. What is the momentum conjugate to t? What is the variational principle written in terms of the extended Hamiltonian K? (Give the explicit form of K and give Hamilton's equations in this form.)

8.13. Consider the double-pendulum problem shown in the figure.

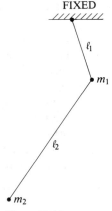

FIXED

ℓ_1

m_1

ℓ_2

m_2

Figure P8.13

(a) Suppose that all the motion is in one plane and introduce appropriate coordinates. What is the dimension of the configuration space? What is the dimension of the phase space $\mathcal{P}$? What is the topology of the phase space (i.e., is it a line, a plane, a sphere, or only part of one of these, or some other topological figure)?

(b) Write down the Lagrangian for this system in a uniform gravitational field. What are the Euler-Lagrange equations in this case?

(c) Find the canonical momenta for this system and write the Hamiltonian H. Is H conserved? Write the Hamilton equations for this system.

8.14. Verify that the determinant of Eq. (8.93) vanishes. [*Hint:* Pick a particular velocity. Rotate coordinates so that this velocity is along the z-axis. Note that the determinant is invariant under this rotation.]

8.15. A particle of mass m slides without friction on an inclined plane in a uniform gravitational field.

(a) Write the Lagrangian.

(b) Obtain the Lagrange equations.

(c) Write the Hamiltonian. Is it conserved?

(d) Obtain the Hamilton equations.

9

***

POISSON BRACKETS

***

9.1 Poisson Brackets and the Equations of Motion

We saw that the Hamilton equations can be written as terms of the $2n$ coordinates $\{z^\mu\}$ of phase space $\mathcal{P}$. If H has explicit time dependence, it is most proper to use the $2n + 1$-dimensional space $\mathcal{P} \times \mathbf{R}$, which has coordinates $\{q^i, p_i, t\}$. The path of the system through this phase spacetime is $q^i(\tau), p_i(\tau), t(\tau)$, where τ is the path parameter defined by $t(\tau) = \tau$. The Hamilton equations then read

$$\frac{\partial H}{\partial z^\mu} = \omega_{\mu\nu}\, z'^\nu \qquad (\mu, \nu \text{ indices range over } 1, \dots, 2n)\,, \qquad (9.1)$$

$$t' = 1 \qquad ('\text{ means } d/d\tau, \text{ but } \tau = t)\,. \qquad (9.2)$$

Thus time is treated differently from the z^μ. The last of the Hamilton equations,

$$\frac{\partial H}{\partial t} = H'\,, \qquad (9.3)$$

is seen not as an equation of motion (in the sense that H is not a coordinate on $\mathcal{P} \times \mathbf{R}$) but as an evolution equation for H.

As a unified notation for the coordinates in phase spacetime, we use $\{z^\mu\}$, where now μ ranges from 0 to $2n + 1$:

$$
\begin{aligned}
z^0 &= t, \\
z^\mu &= q^\mu && \text{for} \quad \mu = 1, \dots, n, \\
z^\mu &= p^{\mu-n} && \text{for} \quad \mu = n+1, \dots, 2n.
\end{aligned}
\tag{9.4}
$$

In many respects this notation is useful for cases when $\mathcal{P}$ is used (then z^0 is not needed) or when $\mathcal{P} \times \mathbf{R}$ is used. We define $\omega_{\mu\nu}$ as before:

$$
\omega_{\mu\nu} = -\omega_{\mu\nu}, \quad \omega_{i,n+i} = -\omega_{n+i,i} = -1 \quad (i = 1, \dots, n),
$$
$$
\text{rest of } \omega_{\mu\nu} = 0.
\tag{9.5}
$$

Notice that when μ or ν has the value 0, $\omega_{\mu\nu} = 0$.

We also define a tensor $\tilde{\omega}$ with upper indices. Its components $\tilde{\omega}^{\mu\nu}$ are defined by

$$
\tilde{\omega}^{\mu\nu} = \begin{cases}
1 & \text{if } \mu = i, \nu = n+i, \\
-1 & \text{if } \nu = i, \mu = n+i \text{ for } i = 1, \dots, n, \\
0 & \text{otherwise.}
\end{cases}
\tag{9.6}
$$

Consequently,

$$
\tilde{\omega}^{\mu\sigma} \omega_{\sigma\nu} = \delta^\mu{}_\nu \text{ if } \mu, \nu = 1, \dots, 2n \text{ and is } 0 \text{ otherwise.}
\tag{9.7}
$$

Caution: If you think of $\omega_{\mu\nu}$ as a matrix, remember that in $\mathcal{P} \times \mathbf{R}$ its determinant is zero. Then $\tilde{\omega}^{\mu\nu}$ is not the inverse of $\omega_{\mu\nu}$, since the "matrix product" $\tilde{\omega}^{\mu\sigma} \omega_{\sigma\nu}$ is $\delta^\mu{}_\nu$ only if neither μ nor ν takes the value 0. On the other hand, in case there is no explicit time dependence and we work with phase space $\mathcal{P}$ by itself, then $\omega_{\mu\nu}$ as a matrix is nonsingular, and $\tilde{\omega}^{\mu\nu}$ is indeed its inverse.

In any case, we can write

$$
z'^\mu = \tilde{\omega}^{\mu\nu} H_{,\nu}, \qquad \mu = 1, \dots, 2n.
\tag{9.8}
$$

Any function $F(q, p, t)$ on phase spacetime becomes a function of τ on the path $q^i(\tau), p_i(\tau), t(\tau)$. Its derivative along the path is

$$
\frac{dF}{d\tau} = \frac{\partial F}{\partial q^i} \frac{dq^i}{d\tau} + \frac{\partial F}{\partial p_i} \frac{dp_i}{d\tau} + \frac{\partial F}{\partial t} \frac{dt}{d\tau}.
\tag{9.9}
$$

An alternative way of writing this derivative is

$$
\frac{dF}{d\tau} = \sum_{\mu=0}^{2n} F_{,\mu} z'^\mu = \sum_{\mu=1}^{2n} F_{,\mu} z'^\mu + \frac{\partial F}{\partial z^0} z'^0.
\tag{9.10}
$$

Since $t' = z'^0 = 1$, we have (now writing $t = \tau$)

$$
\frac{dF}{dt} = \sum_{\mu=1}^{2n} F_{,\mu} \dot{z}^\mu + \frac{\partial F}{\partial t} = F_{,\mu} \tilde{\omega}^{\mu\nu} H_{,\nu} + \frac{\partial F}{\partial t}.
\tag{9.11}
$$

Notice that the summation convention may now be employed unambiguously in the first term, since $\tilde{\omega}^{\mu\nu}$ has nonzero terms only when neither μ nor ν is 0.

The **Poisson bracket** $[\cdot, \cdot]$ is defined for any two functions F, G of $\mathcal{P} \times \mathbf{R}$:

$$[F, G] \equiv F_{,\mu} \, \tilde{\omega}^{\mu\nu} G_{,\nu}. \tag{9.12}$$

Notice that $F = F(q, p, t)$ and $G = G(q, p, t)$, but the t-derivatives are ignored. In terms of q^i and p_i derivatives, the Poisson bracket is

$$[F, G] = \frac{\partial F}{\partial q^i} \frac{\partial G}{\partial p_i} - \frac{\partial F}{\partial p_i} \frac{\partial G}{\partial q^i}. \tag{9.13}$$

The time derivative of a function F may thus be written

$$\frac{dF}{dt} = [F, H] + \frac{\partial F}{\partial t}. \tag{9.14}$$

For special functions the Poisson bracket has special significance. First, consider z^μ, now considered as a function on $\mathcal{P} \times \mathbf{R}$. We have

$$\dot{z}^\mu = [z^\mu, H] + \frac{\partial z^\mu}{\partial t}, \tag{9.15}$$

but the latter term is zero, since z^μ and t are independent variables if $\mu = 1, \dots, 2n$. Consequently,

$$\dot{z}^\mu = [z^\mu, H] \qquad \text{if} \quad \mu = 1, \dots, 2n. \tag{9.16}$$

This form of the Hamilton equations is easily seen to be the same as our previous forms, as in (9.1). We use

$$z^\mu{}_{,\nu} = \delta^\mu{}_\nu \qquad (\mu, \nu \text{ here can range from } 0 \text{ to } 2n) \tag{9.17}$$

to write

$$\dot{z}^\mu = z^\mu{}_{,\alpha} \, \tilde{\omega}^{\alpha\beta} H_{,\beta}, \qquad \mu = 1, \dots, 2n. \tag{9.18}$$

Next, consider t itself, not as a path parameter but as a function on $\mathcal{P} \times \mathbf{R}$:

$$\dot{t} = [t, H] + \frac{\partial t}{\partial t}. \tag{9.19}$$

In this case it is the Poisson bracket, which ignores t-derivatives, that vanishes, and we have

$$\dot{t} = \frac{\partial t}{\partial t} = 1. \tag{9.20}$$

Finally, if the function is H itself, we find that

$$\dot{H} = [H, H] + \frac{\partial H}{\partial t} = \frac{\partial H}{\partial t}. \tag{9.21}$$

In Eq. (9.21), we used the antisymmetry of $[\cdot, \cdot]$:

$$[F, G] = -[G, F], \tag{9.22}$$

so that $[F, F] = 0$.

There is another important property of $[\cdot, \cdot]$: The **Jacobi identity** is

$$\big[[F, G], E\big] + \big[[G, E], F\big] + \big[[E, F], G\big] = 0 \tag{9.23}$$

for any three functions. The proof is simple; first write

$$\begin{aligned}
[[F, G], E] &= (F_{,\mu}\, \tilde{\omega}^{\mu\nu} G_{,\nu})_{,\alpha}\, \tilde{\omega}^{\alpha\beta} E_{,\beta} \\
&= (F_{,\mu\alpha}\, G_{,\nu}\, E_{,\beta} + F_{,\mu}\, G_{,\nu\alpha}\, E_{,\beta})\tilde{\omega}^{\mu\nu}\, \tilde{\omega}^{\alpha\beta}.
\end{aligned} \tag{9.24}$$

Then rewrite this term as

$$[[F, G], E] = (F_{,\alpha\beta}\, G_{,\sigma}\, E_{,\tau} - G_{,\alpha\beta}\, F_{,\sigma}\, E_{,\tau})\tilde{\omega}^{\alpha\sigma}\, \tilde{\omega}^{\beta\tau}. \tag{9.25}$$

(Note that this form is obtained by relabeling the dummy indices, together with exchanging the order of the antisymmetric object $\tilde{\omega}^{\alpha\sigma}$ in the last term.) Then write the other two terms in the Jacobi sum in a similar form; the cancellation by pairs is then obvious.

The Poisson bracket is the basic structure of the Hamilton equations that allows the subject of canonical transformations to be well defined. The Poisson bracket, which is antisymmetric and obeys the Jacobi identity, is important for group theory in Classical Mechanics. The Poisson bracket has many similarities to the commutator of two vector fields. It is important to recognize both similarities and differences between the two concepts, the one defined on functions and the other on vectors. Since the similarities are more provocative than the differences, we have deliberately used $[\cdot, \cdot]$ for both.

To show the connection, we now define a contravariant vector $\mathbf{U}$, with components U^α, defined in the space tangent to phase space by any function $u(q, p)$:

$$\mathbf{U} = u_{,\alpha}\, \tilde{\omega}^{\alpha\beta}\boldsymbol{\partial}_\beta = U^\beta\boldsymbol{\partial}_\beta, \tag{9.26}$$

where the coordinates can be noncanonical if $\tilde{\omega}^{\alpha\beta}$ is suitably transformed. We note that if the vector $\mathbf{U}$ acts on a function $v(q, p)$ on phase space, we have

$$\begin{aligned}
\mathbf{U}(v) &= u_{,\alpha}\, \tilde{\omega}^{\alpha\beta}v_{,\beta} \\
&= [u, v].
\end{aligned} \tag{9.27}$$

Obviously,

$$\mathbf{U}(v) = -\mathbf{V}(u). \tag{9.28}$$

Suppose that we form a second contravariant vector $\mathbf{V} = v_{,\alpha}\, \tilde{\omega}^{\alpha\beta}\boldsymbol{\partial}_\beta = V^\alpha\boldsymbol{\partial}_\alpha$. We may investigate the action of the commutator of $\mathbf{U}$ with $\mathbf{V}$ on a particular phase space function:

$$[\mathbf{U}, \mathbf{V}](f) = \mathbf{U}\big(\mathbf{V}(f)\big) - \mathbf{V}\big(\mathbf{U}(f)\big). \tag{9.29}$$

We note first that the operations are well defined, since the result of a vector operating on a function is another function. Further, we have

$$[\mathbf{U}, \mathbf{V}]f = [\mathbf{U}, \mathbf{V}]^\sigma\boldsymbol{\partial}_\sigma f. \tag{9.30}$$

The point of this equation is that the commutator of two vectors is itself a vector. (Its action on functions is that of a linear combination of first derivatives.) We have

$$[\mathbf{U}, \mathbf{V}]^\sigma = (U^\alpha \boldsymbol{\partial}_\alpha V^\sigma - V^\alpha \boldsymbol{\partial}_\alpha U^\sigma). \tag{9.31}$$

If we return to the definitions of the components U^α, V^α, then

$$
\begin{aligned}
[\mathbf{U}, \mathbf{V}]^\sigma &= (u_{,\gamma} \tilde{\omega}^{\gamma\alpha} \boldsymbol{\partial}_\alpha)(v_{,\mu} \tilde{\omega}^{\mu\sigma}) - (v_{,\gamma} \tilde{\omega}^{\gamma\alpha} \boldsymbol{\partial}_\alpha)(u_{,\mu} \tilde{\omega}^{\mu\sigma}) \\
&= \boldsymbol{\partial}_\mu(u_{,\gamma} \tilde{\omega}^{\gamma\alpha} v_{,\alpha}) \tilde{\omega}^{\mu\sigma} \\
&= [u, v]_{,\mu} \tilde{\omega}^{\mu\sigma}.
\end{aligned}
\tag{9.32}
$$

We have the result that the vector associated with the Poisson bracket is the same as the commutator of the vectors associated with the functions entering the Poisson bracket. Thus the discussion leading to (9.23) is a direct consequence of the Jacobi identity for vectors.

Finally, consider the Poisson bracket of two coordinate functions z^μ and z^ν:

$$[z^\mu, z^\nu] = \tilde{\omega}^{\mu\nu} \qquad (\mu, \nu = 0, \dots, 2n). \tag{9.33}$$

These are the fundamental Poisson brackets and will be used to define canonical coordinates. However, notice that since the Poisson bracket ignores time derivatives, care must be taken to delineate the role of time accurately.

9.2   Forms and Integration

The basic structure of the Hamilton equations is that the gradient $H_{,\mu}$ of the Hamiltonian determines the velocity of the actual path through phase space by use of the tensor $\boldsymbol{\omega}$. Canonical transformations are changes of phase space variables that preserve this structure. As a preliminary, we define some useful concepts concerning antisymmetric tensors, of which $\boldsymbol{\omega}$ is an example, and integration theory.

Briefly an **r-form** is an r-rank tensor whose components have only lower indices and that is completely antisymmetric. An **0-form** is a function F (see Chapter 2). Although we will later work with phase spacetime, we here treat as independent variables z^μ, $\mu = 1, \dots, 2n$. Our consideration will actually apply to any manifold. We will not write indices on z^μ when it is used as an argument in a function as in $F(z)$.

A **1-form** $\boldsymbol{\alpha}$ has been previously defined by

$$\boldsymbol{\alpha} = \alpha_\mu \, \mathbf{d}z^\mu. \tag{9.34}$$

The $\mathbf{d}z^\mu$ have been defined as operators, with the result that $\boldsymbol{\alpha}$ should be thought of as an operator that can be applied to vectors. Here we will develop an interpretation of $\mathbf{d}$ as a differential operator, the **curl** operator, with the result that the inverse of differentiation will be seen to be integration.

Let $\mathcal{C}$ be a curve in phase space parameterized by τ (not necessarily time but any path parameter). In coordinates the path is $z^\mu(\tau)$. The parameter runs monotonically from τ_1 to τ_2. We call P_1 and P_2 the endpoints of the path; their coordinates are $z^\mu(\tau_1)$ and $z^\mu(\tau_2)$. As the parameter varies from τ to $\tau + \delta\tau$, $z(\tau)$ changes from z^μ to $z^\mu + \delta z^\mu$, where

$$\delta z^\mu = \dot{z}^\mu \delta\tau, \qquad \text{where} \quad \dot{z}^\mu = \frac{dz^\mu}{d\tau}. \tag{9.35}$$

This calculation was used previously as the rationale for calling $d/d\tau$ the tangent vector of the path $d/d\tau = \dot{z}^\mu \partial/\partial z^\mu$. We can, in the same spirit, write

$$\mathbf{d}z^\mu = \dot{z}^\mu \mathbf{d}\tau. \tag{9.36}$$

Now we can express $\boldsymbol{\alpha}$, on the path, as $\mathbf{d}\tau$ times a function of τ:

$$\boldsymbol{\alpha} = \alpha_\mu \mathbf{d}z^\mu = \alpha_\mu \dot{z}^\mu \mathbf{d}\tau. \tag{9.37}$$

This expression allows us to define the integral of $\boldsymbol{\alpha}$ along the path $\mathcal{C}$:

$$\int_{P_1,\mathcal{C}}^{P_2} \boldsymbol{\alpha} \equiv \int_{\tau_1}^{\tau_2} \alpha_\mu \dot{z}^\mu \mathbf{d}\tau. \tag{9.38}$$

The left side looks a trifle strange, since the $\mathbf{d}z^\mu$'s are hidden in $\boldsymbol{\alpha}$ and are not explicitly written. The right side, however, is just the ordinary line integral of $\boldsymbol{\alpha}$ along $\mathcal{C}$.

We also had previously defined $\mathbf{d}F$ as the 1-form

$$\mathbf{d}F = F_{,\mu}\, \mathbf{d}z^\mu, \qquad \text{where} \quad F_{,\mu} = \frac{\partial F}{\partial z^\mu}.$$

We can integrate this 1-form along $\mathcal{C}$ to get:

Fundamental Theorem of Calculus _____

The integral of $\mathbf{d}F$ *along a curve* $\mathcal{C}$ *depends only on the values* F_1 *and* F_2 *of the function* F *at the endpoints and is given by*

$$\int_{P_1,\mathcal{C}}^{P_2} \mathbf{d}F = F_2 - F_1. \tag{9.39}$$

_____

The proof of this theorem comes directly from the definition of the line integral:

$$\int_{P_1,\mathcal{C}}^{P_2} \mathbf{d}F = \int_{\tau_1}^{\tau_2} F_{,\mu} \dot{z}^\mu \mathbf{d}\tau = \int_{\tau_1}^{\tau_2} \frac{dF}{d\tau} \mathbf{d}\tau = F_2 - F_1. \tag{9.40}$$

Notice that the integral of $\mathbf{d}F$ about any closed curve [one with $z^\mu(\tau_1) = z^\mu(\tau_2)$] vanishes, since $F_1 = F_2$.

Conversely, we have the important theorem:

Integrability Theorem for 1-Forms_____

Suppose that $\int_C \alpha = 0$ for any closed curve C. Then there is a function F such that $\alpha = \mathbf{d}F$.

_____

The proof starts with the choice of a reference point P_1 and any second point P with coordinates z^μ. We pick an arbitrary value F_1 for the value of F at P_1. We do assume that there is at least a curve joining P_1 to P, and we define $F(z)$ by

$$F(z) = F_1 + \int_{P_1,C}^{P} \alpha. \tag{9.41}$$

This integral is independent of the curve C. To show this independence, consider the integral of α along a second path C^* joining P_1 to P. This second path uses a path parameter τ^*, and if we change to $-\tau^*$, we get the negative path $-C^*$, which goes from P to P_1. We can adjoin $-C^*$ to C by adjusting their respective path parameters to form a closed path with a single path parameter. The integral of α along this path is zero. We now define $F^*(z)$ by using the integral along C^* and using the same assigned F_1:

$$F^*(z) = \int_{P_1,C^*}^{P} \alpha + F_1. \tag{9.42}$$

Because the integral of α along the combined path $C - C^*$ vanishes, the values of the integrals in (9.41) and (9.42) are equal; therefore,

$$F^* = F. \tag{9.43}$$

In other words, $F(z)$ is independent of path.

To find $\mathbf{d}F$, we must compute $\partial F/\partial z^\mu$. We extend C for a small value of τ to $\tau + \delta\tau$, where the coordinates are $z^\mu + \delta z^\mu$. The change in F is

$$\delta F = \int_{\tau_1}^{\tau+\delta\tau} \alpha_\mu \dot{z}^\mu \, d\tau - \int_{\tau_1}^{\tau} \alpha_\mu \dot{z}^\mu \, d\tau = \alpha_\mu \dot{z}^\mu \delta\tau = \alpha_\mu \delta z^\mu. \tag{9.44}$$

This calculation is just the one we need to calculate $\partial F/\partial z^\mu$, and the result may therefore be interpreted directly in terms of 1-forms:

$$\mathbf{d}F = \alpha_\mu \, \mathbf{d}z^\mu, \tag{9.45}$$

as desired.

A **2-form** is an antisymmetric, covariant rank 2 tensor. Like any other rank 2 covariant tensor, it can be thought of as a bilinear operator acting on vectors to yield a function. Let the 2-form be denoted $\mathbf{F}$ and let $\mathbf{U}, \mathbf{V}$ be any two vectors. Then

$$\mathbf{F}(\mathbf{U}, \mathbf{V}) = \phi \tag{9.46}$$

is a function. $\mathbf{F}$ is linear in each argument. Thus, if $\mathbf{U}_1$ and $\mathbf{U}_2$ are vectors and if α_1 and α_2 are functions, then

$$\mathbf{F}(\alpha_1\mathbf{U}_1 + \alpha_2\mathbf{U}_2, \mathbf{V}) = \alpha_1\mathbf{F}(\mathbf{U}_1, \mathbf{V}) + \alpha_2\mathbf{F}(\mathbf{U}_2, \mathbf{V}). \tag{9.47}$$

Furthermore, $\mathbf{F}$ is antisymmetric:

$$\mathbf{F}(\mathbf{U}, \mathbf{V}) = -\mathbf{F}(\mathbf{V}, \mathbf{U}). \tag{9.48}$$

Just as the general 1-form is a linear combination of the elementary 1-forms $\mathbf{d}z^\mu$, so $\mathbf{F}$ may be written as a sum of quadratic terms involving the $\mathbf{d}z^\mu$. The symbol $\otimes$ is used to denote tensor product. Here, however, we will use the **wedge** symbol $\wedge$ to denote an antisymmetric tensor product:

$$\begin{aligned} \mathbf{d}z^\mu \wedge \mathbf{d}z^\nu &= \tfrac{1}{2}(\mathbf{d}z^\mu \otimes \mathbf{d}z^\nu - \mathbf{d}z^\nu \otimes \mathbf{d}z^\mu) \\ &= -\mathbf{d}z^\nu \wedge \mathbf{d}z^\mu. \end{aligned} \tag{9.49}$$

The definition of this product is given by showing how an elementary 2-form $\mathbf{d}z^\mu \wedge \mathbf{d}z^\nu$ acts on two vectors $\mathbf{U} = u^\mu \boldsymbol{\partial}_\mu$ and $\mathbf{v} = v^\mu \boldsymbol{\partial}_\mu$:

$$\mathbf{d}z^\mu \wedge \mathbf{d}z^\nu(\mathbf{U}, \mathbf{V}) = \tfrac{1}{2}(u^\mu v^\nu - v^\mu u^\nu). \tag{9.50}$$

The components $F_{\mu\nu}$ of a general $\mathbf{F}$ are defined by

$$\mathbf{F} = \tfrac{1}{2}F_{\mu\nu}\,\mathbf{d}z^\mu \wedge \mathbf{d}z^\nu, \qquad \text{where} \quad F_{\mu\nu} = -F_{\mu\nu}. \tag{9.51}$$

Consequently, we have

$$\mathbf{F}(\mathbf{U}, \mathbf{V}) = \tfrac{1}{2}F_{\mu\nu}\tfrac{1}{2}(u^\mu v^\nu - v^\mu u^\nu) = \tfrac{1}{2}F_{\mu\nu}\,u^\mu v^\nu. \tag{9.52}$$

We can also write

$$\mathbf{F} = \sum_{\mu<\nu} F_{\mu\nu}\,\mathbf{d}z^\mu \wedge \mathbf{d}z^\nu, \tag{9.53}$$

where the sum indicates that only terms with $\mu < \nu$ are included. Of course, terms with $\mu = \nu$ vanish anyway.

Consider, for example, a two-dimensional space, with coordinates $\{y^1, y^2\}$. The general 2-form is

$$\mathbf{F} = \tfrac{1}{2}F_{\mu\nu}\,\mathbf{d}y^\mu \wedge \mathbf{d}y^\nu = F_{12}\,\mathbf{d}y^1 \wedge \mathbf{d}y^2. \tag{9.54}$$

If we change variables to $\bar{y}^\mu$:

$$\bar{y}^\mu = \bar{y}^\mu(y) \qquad \text{or} \qquad y^\mu = y^\mu(\bar{y}), \tag{9.55}$$

we compute the change in the components of $\mathbf{F}$ as follows:

$$\begin{aligned} \mathbf{F} &= \tfrac{1}{2}F_{\mu\nu}\,\mathbf{d}y^\mu \wedge \mathbf{d}y^\nu = \tfrac{1}{2}F_{\bar{\mu}\bar{\nu}}\,\mathbf{d}\bar{y}^\mu \wedge \mathbf{d}\bar{y}^\nu \\ &= \tfrac{1}{2}F_{\bar{\mu}\bar{\nu}}\frac{\partial\bar{y}^\mu}{\partial y^\alpha}\frac{\partial\bar{y}^\nu}{\partial y^\beta}\,\mathbf{d}y^\alpha \wedge \mathbf{d}y^\beta. \end{aligned} \tag{9.56}$$

The $F_{\bar{\mu}\bar{\nu}}$ are the components of $\mathbf{F}$ in the new coordinates. Clearly, $F_{\bar{\mu}\bar{\nu}} = -F_{\bar{\nu}\bar{\mu}}$ and in two dimensions there is just one independent component:

$$\bar{F} = F_{12}\,\mathbf{d}y^1 \wedge \mathbf{d}y^2 = F_{\bar{1}\bar{2}}\left(\frac{\partial \bar{y}^1}{\partial y^1}\frac{\partial \bar{y}^2}{\partial y^2} - \frac{\partial \bar{y}^2}{\partial y^1}\frac{\partial \bar{y}^1}{\partial y^2}\right)\mathbf{d}y^1 \wedge \mathbf{d}y^2 \tag{9.57}$$

$$= F_{\bar{1}\bar{2}}\,\mathbf{d}\bar{y}^1 \wedge \mathbf{d}\bar{y}^2.$$

Notice that in this formula appears the **Jacobian** of the transformation, that is, the determinant of the 2×2 matrix $\partial \bar{y}^\mu / \partial y^\alpha$.

The integral of $\mathbf{F}$ over a region R in the two-dimensional space is defined to be

$$\int_R \mathbf{F} = \int_R F_{12}\,\mathbf{d}y^1\,\mathbf{d}y^2. \tag{9.58}$$

This definition works for any coordinates:

$$\int_R \mathbf{F} = \int_R F_{\bar{1}\bar{2}}\,\mathbf{d}\bar{y}^1\,\mathbf{d}\bar{y}^2, \tag{9.59}$$

since the transformation law for $F_{12} \to F_{\bar{1}\bar{2}}$ includes the Jacobian factor needed in making the transition $\mathbf{d}y^1\,\mathbf{d}y^2 \to \mathbf{d}\bar{y}^1\,\mathbf{d}\bar{y}^2$.

There is one cautionary word: The Jacobian must not be zero (or else the coordinate transformation is singular), but it may be positive or it may be negative. In the latter case, it is often the practice to use the absolute value of the Jacobian in defining the integral in the new coordinates. We will not do so; to emphasize that in fact we use the Jacobian itself, positive or negative, we write the "volume" element not as $dy^1\,dy^2$ but as

$$\mathbf{d}y^1 \wedge \mathbf{d}y^2. \tag{9.60}$$

(To compute the integral, it is helpful to write $dy^1\,dy^2$ and use ordinary integral techniques. Be very careful of the overall sign, though.)

A 2-form $\mathbf{F}$ in a higher-dimensional space is

$$\mathbf{F} = \tfrac{1}{2} F_{\mu\nu}\,\mathbf{d}z^\nu \wedge \mathbf{d}z^\nu. \tag{9.61}$$

In this space a two-dimensional subspace is defined by giving z^μ as a function of two variables y^1, y^2. Let us denote by y^A, $A = 1, 2$, these two variables; thus

$$z^\mu = z^\mu(y) \tag{9.62}$$

denotes the surface or subspace. The integral of $\mathbf{F}$ over a subset R of this surface is defined by

$$\int_R \mathbf{F} = \int_{R_y} \frac{1}{2} F_{\mu\nu}\left(\frac{\partial z^\mu}{\partial y^A}\right)\left(\frac{\partial z^\nu}{\partial y^B}\right)\mathbf{d}y^A \wedge \mathbf{d}y^B, \tag{9.63}$$

where R_y represents the region in $(y^1\text{–}y^2)$-space which corresponds to R, the subset of the 2-surface in z^μ space. This integral may be written in the forms

$$\int_R \mathbf{F} = \int_{R_y} \frac{1}{2} F_{\mu\nu}\left(\frac{\partial z^\mu}{\partial y^1}\frac{\partial z^\nu}{\partial y^2} - \frac{\partial z^\mu}{\partial y^2}\frac{\partial z^\nu}{\partial y^1}\right)\mathbf{d}y^1 \wedge \mathbf{d}y^2, \tag{9.64}$$

$$= \int_{R_y} F_{\mu\nu} \frac{\partial z^\mu}{\partial y^1} \frac{\partial z^\nu}{\partial y^2} \, \mathbf{d}y^1 \wedge \mathbf{d}y^2. \tag{9.65}$$

Suppose that **F** has the special form

$$\mathbf{F} = \alpha_{\mu,\nu} \, \mathbf{d}z^\nu \wedge \mathbf{d}z^\mu. \tag{9.66}$$

Notice the order of the indices and the lack of a $\frac{1}{2}$:

$$\mathbf{F} = \tfrac{1}{2} F^{\mu\nu} \, \mathbf{d}z^\nu \wedge \mathbf{d}z^\mu, \qquad \text{where} \quad F_{\mu\nu} = \alpha_{\nu,\mu} - \alpha_{\mu,\nu}. \tag{9.67}$$

Further, suppose that R_y is defined by the rectangle $a_1 < y^1 < b_1$, $a_2 < y^2 < b_2$, where a_1, a_2, b_1, b_2 are constants. We then have, from (9.65),

$$\int_R \mathbf{F} = \int_{a_1}^{b_1} \int_{a_2}^{b_2} [\alpha_{\nu,\mu} - \alpha_{\mu,\nu}] \frac{\partial z^\mu}{\partial y^1} \frac{\partial z^\nu}{\partial y^2} \, \mathbf{d}y^1 \wedge \mathbf{d}y^2. \tag{9.68}$$

We note that the α_μ are functions of z^μ and therefore of y^A. Therefore, $\partial \alpha_\mu / \partial y^A$ is defined, and we have

$$\int_R \mathbf{F} = \int_{a_1}^{b_1} \int_{a_2}^{b_2} \left(\frac{\partial \alpha_\nu}{\partial y^1} \frac{\partial z^\nu}{\partial y^2} - \frac{\partial \alpha_\nu}{\partial y^2} \frac{\partial z^\nu}{\partial y^1} \right) \mathbf{d}y^1 \wedge \mathbf{d}y^2. \tag{9.69}$$

We use now the fact that

$$\frac{\partial^2 z^\mu}{\partial y^1 \, \partial y^2} = \frac{\partial^2 z^\mu}{\partial y^2 \, \partial y^1} \tag{9.70}$$

to write this integral as

$$\int_R \mathbf{F} = \int_{a_1}^{b_1} \int_{a_2}^{b_2} \left[\frac{\partial}{\partial y^1} \left(\alpha_\nu \frac{\partial z^\nu}{\partial y^2} \right) - \frac{\partial}{\partial y^2} \left(\alpha_\nu \frac{\partial z^\nu}{\partial y^1} \right) \right] \mathbf{d}y^1 \wedge \mathbf{d}y^2. \tag{9.71}$$

Each term may be integrated once:

$$\int_R \mathbf{F} = \int_{a_2}^{b_2} \alpha_\nu \frac{\partial z^\nu}{\partial y^2} \, dy^2 \Big|_{a_1}^{b_1} - \int_{a_1}^{b_1} \alpha_\nu \frac{\partial z^\nu}{\partial y^1} \, dy^1 \Big|_{a_2}^{b_2}. \tag{9.72}$$

To interpret the right side of this expression, remember that the boundary of R is a closed curve defined by the function $z^\mu(y)$ on the boundary of R_y. We denote the boundary of R by ∂R. The first term, of course, is shorthand for the difference

$$\int_{a_2}^{b_2} \alpha_\nu \frac{\partial z^\nu}{\partial y^2} \, dy^2 \Big|_{y^1 = b_1} - \int_{a_2}^{b_2} \alpha_\nu \frac{\partial z^\nu}{\partial y^2} \, dy^2 \Big|_{y^1 = a_1}. \tag{9.73}$$

The first of these two integrals is just the line integral of $\alpha_\mu \, dz^\mu$ along that part of ∂R defined by $y^1 = b_1$ (i.e., ∂R_2 in Figure 9.1.) The second integral is the negative of the line integral along ∂R^4 defined by $y^1 = a_1$, but the sign is reversed if we choose an oppositely directed path parameter. We therefore choose as path parameter along ∂R_2 the parameter $-y^2$. We see that the rest of the expression for $\int_R \mathbf{F}$ is also a pair of line integrals, on ∂R_3 and ∂R_1.

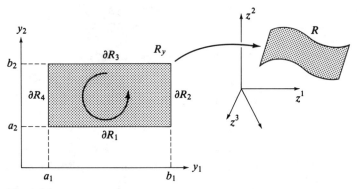

Figure 9.1

Moreover, the signs of these integrals are such that one consistent direction along all of ∂R may be chosen. This direction is the "circulation" direction given by the right-hand rule. (Place your right hand with fingers along the y^1-axis and palm facing the y^2-axis. Your fingers curl in the correct direction. This convention, of course, is just that, a convention. If we had drawn a mirror image of the figure, we would have needed a left-hand rule. In other words, be careful.)

In summary, the right side of $\int_R \mathbf{F}$ consists of four line segment integrals along the four parts of ∂R. The integrands are all $\alpha_\mu \, dz^\mu$ suitably transformed into something like $\alpha_\mu (\partial z^\mu / \partial y) \, dy$. We therefore write $\boldsymbol{\alpha} = \alpha_\mu \, \mathbf{d}z^\mu$, and we have

$$\int_R \mathbf{F} = \int_{\partial R} \boldsymbol{\alpha}. \tag{9.74}$$

The 2-form $\mathbf{F}$ was obtained from the 1-form $\boldsymbol{\alpha}$ by differentiation. We define the operator $\mathbf{d}$ acting on 1-forms such as $\boldsymbol{\alpha}$ by

$$\mathbf{d}\boldsymbol{\alpha} = \mathbf{d}(\alpha_\mu \, \mathbf{d}z^\mu) = \alpha_{\mu,\nu} \, \mathbf{d}z^\nu \wedge \mathbf{d}z^\mu. \tag{9.75}$$

The result for the integral of $\mathbf{F}$ is thus best written by emphasizing that $\mathbf{F} = \mathbf{d}\boldsymbol{\alpha}$:

$$\int_R \mathbf{d}\boldsymbol{\alpha} = \int_{\partial R} \boldsymbol{\alpha} \qquad \text{for} \quad R \text{ a simply connected region,} \tag{9.76}$$

and we have added the note to indicate that we have only proved the theorem for a relatively simple region.

This formula works for all forms; we now define the general r-**form** and the action of $\mathbf{d}$ on it. First, the antisymmetric multiplication **wedge** product $\wedge$ is extended to a general term of the form

$$\mathbf{d}z^\alpha \wedge \mathbf{d}z^\beta \wedge \cdots \wedge \mathbf{d}z^\lambda \qquad (r \text{ terms}). \tag{9.77}$$

This term is a tensor of rank r: It is a multilinear operator that acts on r vectors $\mathbf{A}, \mathbf{B}, \cdots, \mathbf{C}$ (r factors) having components $a^\alpha, b^\alpha, \ldots, c^\alpha$. Since $\wedge$ is antisymmetric, the r-form of (9.77) is completely antisymmetric. The action on $\mathbf{A}, \mathbf{B}, \ldots, \mathbf{C}$ is to produce a function that changes sign whenever any two vectors are interchanged. The result is

$$\mathbf{d}z^\alpha \wedge \mathbf{d}z^\beta \wedge \cdots \wedge \mathbf{d}z^\lambda \, (\mathbf{A}, \mathbf{B}, \ldots, \mathbf{C}) = a^{[\alpha} \, b^\beta \ldots c^{\lambda]},$$

where the brackets around the indices indicate the sum

$$a^{[\alpha} \, b^\beta \ldots c^{\gamma]} = \frac{1}{r!} \sum (-1)^\pi \, a^{\pi(\alpha)} \, b^{\pi(\beta)} \ldots c^{\pi(\gamma)}, \tag{9.78}$$

where π is a permutation of the indices and where $(-1)^\pi = +1$ if π is an even permutation and $(-1)^\pi = -1$ if π is an odd permutation.

Two examples should suffice to illustrate this definition. First, the general 3-form $\mathbf{D}$ is of the form

$$\mathbf{D} = \frac{1}{3!} D_{\alpha\beta\gamma} \, \mathbf{d}z^\alpha \wedge \mathbf{d}z^\beta \wedge \mathbf{d}z^\gamma, \tag{9.79}$$

where $D_{\alpha\beta\gamma}$ is a completely antisymmetric array. Each term, such as $\mathbf{d}z^\alpha \wedge \mathbf{d}z^\beta \wedge \mathbf{d}z^\gamma$, has the following effect on three vectors:

$$\mathbf{d}z^\alpha \wedge \mathbf{d}z^\beta \wedge \mathbf{d}z^\gamma (\mathbf{A}, \mathbf{B}, \mathbf{C}) = a^{[\alpha} b^\beta c^{\gamma]}$$
$$= \frac{1}{3!} (a^\alpha b^\beta c^\gamma + a^\beta b^\gamma c^\alpha + a^\gamma b^\alpha c^\beta - a^\gamma b^\beta c^\alpha - a^\beta b^\alpha c^\gamma - a^\alpha b^\gamma c^\beta). \tag{9.80}$$

Therefore, D acting on $\mathbf{A}, \mathbf{B}, \mathbf{C}$ yields

$$\mathbf{D}(\mathbf{A}, \mathbf{B}, \mathbf{C}) = \frac{1}{3!} D_{\alpha\beta\gamma} \, a^{[\alpha} b^\beta c^{\gamma]}$$
$$= \frac{1}{3!} D_{\alpha\beta\gamma} \, a^\alpha b^\beta c^\gamma. \tag{9.81}$$

The last line follows from the complete antisymmetry of $D_{\alpha\beta\gamma}$.

The second example starts from the 2-form $\boldsymbol{\omega} = \mathbf{d}q^i \wedge \mathbf{d}p_i$ in a four-dimensional phase space with coordinates q^1, q^2, p_1, p_2. This 2-form is

$$\boldsymbol{\omega} = \mathbf{d}q^1 \wedge \mathbf{d}p_1 + \mathbf{d}q^2 \wedge \mathbf{d}p_2. \tag{9.82}$$

If we let $z^1 = q^1$, $z^2 = q^2$, $z^3 = p_1$, $z^4 = p_2$, we have

$$\boldsymbol{\omega} = \mathbf{d}z^1 \wedge \mathbf{d}z^3 + \mathbf{d}z^2 \wedge \mathbf{d}z^4. \tag{9.83}$$

The action on vectors $\mathbf{A} = a^\mu \boldsymbol{\partial}_\mu$ and $\mathbf{B} = b^\mu \boldsymbol{\partial}_\mu$ is

$$\boldsymbol{\omega}(\mathbf{A}, \mathbf{B}) = \tfrac{1}{2}(a^1 b^3 - a^3 b^1) + \tfrac{1}{2}(a^2 b^4 - a^4 b^2). \tag{9.84}$$

Further, we can compute $\boldsymbol{\omega} \wedge \boldsymbol{\omega}$ by simply treating $\wedge$ as an associative multiplication. Because of the antisymmetric nature of $\wedge$, the terms add to produce

$$\boldsymbol{\omega} \wedge \boldsymbol{\omega} = -2 \, \mathbf{d}z^1 \wedge \mathbf{d}z^2 \wedge \mathbf{d}z^3 \wedge \mathbf{d}z^4.$$

This result is a 4-form, which acts on quadruples of vectors $\mathbf{A}, \mathbf{B}, \mathbf{C}, \mathbf{D}$ to produce a function that is a sum of 24 ($= 4!$) terms. These terms include $(1/4!)a^1 b^2 c^3 d^4$

and $(1/4!)a^2b^1c^4d^3$ (since $\{2, 1, 4, 3\}$ is an even permutation of $\{1, 2, 3, 4\}$) as well as $(-1/4!)a^2b^3c^4d^1$ (since $\{2, 3, 4, 1\}$ is an odd permutation of $\{1, 2, 3, 4\}$) and, of course, other terms.

The operator **d** (the **curl** operator) can be applied to any form as follows: The general r-form is

$$\mathbf{G} = \frac{1}{r!}\, g_{\alpha\ldots\beta}\, \mathbf{d}z^\alpha \wedge \cdots \wedge \mathbf{d}z^\beta. \tag{9.85}$$

The action of **d** produces an $(r+1)$-form:

$$\mathbf{d}G = \frac{1}{r!}\, g_{\alpha\ldots\beta,\gamma}\, \mathbf{d}z^\gamma \wedge \mathbf{d}z^\alpha \wedge \cdots \wedge \mathbf{d}z^\beta. \tag{9.86}$$

Needless to say, the components $g_{\alpha\ldots\beta}$ of the original r-form form a completely antisymmetric array (or else $\mathbf{d}z^\alpha \wedge \cdots \wedge \mathbf{d}z^\beta$ picks the completely antisymmetric part):

$$g_{\alpha\ldots\beta} = g_{[\alpha\ldots\beta]}. \tag{9.87}$$

The components of $\mathbf{d}G$ are actually the antisymmetric part of the derivative

$$g_{[\alpha\ldots\beta,\gamma]}. \tag{9.88}$$

For example, let **F** be the 2-form

$$\mathbf{F} = \tfrac{1}{2}\, F_{\alpha\beta}\, \mathbf{d}z^\alpha \wedge \mathbf{d}z^\beta. \tag{9.89}$$

Then $\mathbf{d}F$ is

$$\begin{aligned}
\mathbf{d}F &= \tfrac{1}{2}\, F_{\alpha\beta,\gamma}\, \mathbf{d}z^\gamma \wedge \mathbf{d}z^\alpha \wedge \mathbf{d}z^\beta \\
&= \tfrac{1}{2}\, F_{[\alpha\beta,\gamma]}\, \mathbf{d}z^\gamma \wedge \mathbf{d}z^\alpha \wedge \mathbf{d}z^\beta,
\end{aligned} \tag{9.90}$$

where

$$\begin{aligned}
F_{[\alpha\beta,\gamma]} &= \tfrac{1}{6}\left[F_{\alpha\beta,\gamma} + F_{\beta\gamma,\alpha} + F_{\gamma\alpha,\beta} - F_{\beta\alpha,\gamma} - F_{\gamma\beta,\alpha} - F_{\alpha\gamma,\beta} \right] \\
&= \tfrac{1}{3}\left[F_{\alpha\beta,\gamma} + F_{\beta\gamma,\alpha} + F_{\gamma\alpha,\beta} \right].
\end{aligned} \tag{9.91}$$

The simplification to the last expression is due to the antisymmetry of $F_{\alpha\beta}$ itself.

The double application of **d** to any form gives zero. We see this as follows: Let **G** be the r-form as above. Then

$$\mathbf{dd}G = \frac{1}{r!}\, g_{\alpha\ldots\beta,\gamma\delta}\, \mathbf{d}z^\delta \wedge \mathbf{d}z^\gamma \wedge \mathbf{d}z^\alpha \wedge \cdots \wedge \mathbf{d}z^\beta. \tag{9.92}$$

Since second derivatives commute, the result is

$$\mathbf{dd}G = 0. \tag{9.93}$$

As an example, we may consider a covariant 4-vector (a 1-form) given in spacetime by

$$\begin{aligned}
\mathbf{A} &= A_i\, \mathbf{d}x^i - \phi\, \mathbf{d}t \qquad (i = 1,\, 2,\, 3) \\
&= A_\mu\, \mathbf{d}x^\mu \qquad (\mu = 0, \cdots, 3;\ \text{note that}\ x^0 = t),
\end{aligned} \tag{9.94}$$

where A_i are the components of the electromagnetic vector potential $\vec{A}$, and ϕ is the electric scalar potential. Now

$$\mathbf{dA} = A_{[\mu,\sigma]}\,\mathbf{d}z^{\sigma} \wedge \mathbf{d}x^{\mu}. \tag{9.95}$$

Let

$$F_{\sigma\mu} = 2A_{[\mu,\sigma]}; \tag{9.96}$$

then

$$\mathbf{dA} = \mathbf{F}$$
$$= \tfrac{1}{2}\,F_{\sigma\mu}\,\mathbf{d}x^{\sigma} \wedge \mathbf{d}x^{\mu}. \tag{9.97}$$

By working out the components of $F_{\sigma\mu}$, we have, for instance,

$$F_{12} = \partial_1 A_2 - \partial_2 A_1 = (\vec{\nabla} \times \vec{A})_3 = B_3 \tag{9.98}$$

(where $\vec{B}$ is the magnetic vector), and

$$F_{10} = \partial_1 A_0 - \partial_0 A_1 = \dot{A} - (\vec{\nabla}\phi)_1 - (\vec{A})_2' = E_1 \tag{9.99}$$

(where $\vec{E}$ is the electric vector). Hence

$$F_{\mu\nu} = \begin{pmatrix} 0 & -E_1 & -E_2 & -E_3 \\ E_1 & 0 & B_3 & -B_2 \\ E_2 & -B_3 & 0 & B_1 \\ E_3 & B_2 & -B_1 & 0 \end{pmatrix}. \tag{9.100}$$

This **electromagnetic field tensor** is an antisymmetric rank-2 covariant tensor **F**, a 2-form. Maxwell's equations are (for source-free fields)

$$\vec{\nabla} \times \vec{E} + \frac{\partial \vec{B}}{\partial t} = 0, \qquad \vec{\nabla} \cdot \vec{B} = 0, \tag{9.101a}$$

$$\vec{\nabla} \times \vec{B} - \frac{\partial \vec{E}}{\partial t} = 0, \qquad \vec{\nabla} \cdot \vec{E} = 0. \tag{9.101b}$$

The first two equations, when written in terms of the 2-form **F**, read

$$\mathbf{d}F = 0, \tag{9.102}$$

which is, in this context, an identity, since we have already defined $\mathbf{F} = \mathbf{dA}$.

As another example, notice that the fundamental tensor $\boldsymbol{\omega}$ may be obtained by taking the curl of a potential:

$$\mathbf{d}\,(p_i\,\mathbf{d}x^i) = \mathbf{d}p_i \wedge \mathbf{d}x^i. \tag{9.103}$$

An r-form serves as the integrand for an integral over an r-dimensional subset R: $\int_R \mathbf{G}$. If $\mathbf{G} = \mathbf{d}F$ for some $(r-1)$-form **F**, we have the integration theorem

$$\int_R \mathbf{d}F = \int_{\partial R} \mathbf{F} \qquad \text{(for simple } R \text{, at least),} \tag{9.104}$$

where ∂R is the $(r - 1)$-dimensional boundary of R. We will not prove this theorem, but you should try it for the rectangular type of R as we did above for the two-dimensional case.

If $\mathbf{G} = \mathbf{d}F$, then of course $\mathbf{d}G = 0$. A form that satisfies $\mathbf{d}G = 0$ is called a **closed form** or **curl-free**. If also $\mathbf{G} = \mathbf{d}F$ for some F, then G is called an **exact form**. Consider, for example, the one-dimensional space consisting of the unit circle. The 1-form $\mathbf{d}\theta$ (θ = angle measured around the circle) is closed but not exact: $\mathbf{d}\theta$ is defined everywhere on the circle, but it is not the curl of a function, since θ cannot be defined everywhere (i.e., θ cannot be defined as a single-valued functions without jumps). The notation $\mathbf{d}\theta$ is therefore a misleading (albeit a common) one.

We can say, however, that if $\mathbf{d}F = 0$ in a suitably simple region R, then in that region a form $\mathbf{A}$ can be found such that $\mathbf{F} = \mathbf{d}A$. By suitably simple, we mean that R must be *star-shaped* in that at least one point exists (call it z_0) such that each point in R can be connected to it by a "straight line" lying entirely in R. That is, if any point z^μ is in R, then so is $z_0^\mu + \lambda (z^\mu - z_0^\mu)$ for all $0 < \lambda < 1$. (More general regions may be built out of these R's, just as the integration theorems may be implied to complicated sets built up out of the simple sets needed to prove the theorems.) We call the coordinates of the chosen point z_0^μ, and if

$$F = \frac{1}{r!} f_{\alpha \ldots \beta} \, \mathbf{d}z^\alpha \wedge \cdots \wedge \mathbf{d}z^\beta \qquad (9.105)$$

is such that $\mathbf{d}F = 0$, we have

$$f_{[\alpha \ldots \beta, \gamma]} = 0. \qquad (9.106)$$

The form $\mathbf{A} = [1/(r - 1)!] \, a_{\alpha \ldots \beta} \, \mathbf{d}z^\alpha \wedge \mathbf{d}z^\beta$, an $(r - 1)$-form, has components given by the one-dimensional integral

$$a_{\alpha \ldots \beta}(z) = (-1)^{r-1} \int_0^1 \lambda^{r-1} \left(z^\sigma - z_0^\sigma \right) f_{\alpha \ldots \beta \sigma} \left[z_0 + \lambda (z - z_0) \right] d\lambda. \qquad (9.107)$$

Then $\mathbf{F} = \mathbf{d}A$. The constant r is the rank of the form $\mathbf{F}$, and the integral is with respect to the variable λ; the argument of $f_{\alpha \ldots \beta \sigma}$ is in brackets and indicates that $f_{\alpha \ldots \beta \sigma}$ is to be evaluated at the point $z_0^\mu + \lambda (z^\mu - z_0^\mu)$. We will leave as an exercise the computation that $\mathbf{d}A = \mathbf{F}$ except for one illustrative case related to the discussion above of the electromagnetic tensor.

The special case is when $\mathbf{F}$ is a curl-free 2-form,

$$\mathbf{F} = \tfrac{1}{2} F_{\alpha\beta} \, \mathbf{d}z^\alpha \wedge \mathbf{d}z^\beta, \qquad \text{where} \quad F_{[\alpha\beta, \lambda]} = 0.$$

In this case $\mathbf{A}$ is a 1-form

$$\mathbf{A} = A_\alpha \, \mathbf{d}z^\alpha, \qquad \text{where} \quad A_\alpha = -\int_0^1 \lambda \left(z^\sigma - z_0^\sigma \right) F_{\alpha\sigma} \left[z_0 + \lambda (z - z_0) \right] d\lambda.$$

The curl of $\mathbf{A}$ is

$$\mathbf{dA} = A_{\alpha,\beta} \, \mathbf{d}z^\beta \wedge \mathbf{d}z^\alpha = -\tfrac{1}{2}(A_{\alpha,\beta} - A_{\beta,\alpha}) \, \mathbf{d}z^\alpha \wedge \mathbf{d}z^\beta.$$

Therefore, we verify that

$$-A_{\alpha,\beta} + A_{\beta,\alpha} = F_{\alpha\beta}.$$

To do so, we must first compute

$$A_{\alpha,\beta} = -\int_0^1 \{\lambda \, F_{\alpha\beta} \, [z_0 + \lambda \, (z - z_0)] + (z^\sigma - z_0^\sigma)\lambda^2 \, F_{\alpha\sigma,\beta}\} \, d\lambda.$$

Since

$$F_{\alpha\sigma,\beta} + F_{\sigma\beta,\alpha} + F_{\beta\alpha,\sigma} = 0,$$

we have

$$-A_{\alpha,\beta} + A_{\beta,\alpha} = \int_0^1 \lambda[2 \, F_{\alpha\beta} + \lambda \, (z^\sigma - z_0^\sigma) \, F_{\alpha\beta,\sigma}] \, d\lambda.$$

However,

$$\frac{d}{d\lambda}\{F_{\alpha\beta} \, [z_0 + \lambda \, (z - z_0)] \, \} = (z^\sigma - z_0^\sigma) \, F_{\alpha\beta,\sigma}.$$

Consequently,

$$-A_{\alpha,\beta} + A_{\beta,\alpha} = \int_0^1 \left(2\lambda \, F_{\alpha\beta} + \lambda^2 \frac{d}{d\lambda} F_{\alpha\beta}\right) d\lambda$$

$$= \int_0^1 \frac{d}{d\lambda}(\lambda^2 \, F_{\alpha\beta}) \, d\lambda = \lambda^2 \, F_{\alpha\beta}\Big|_0^1 = F_{\alpha\beta} \, (z).$$

Even though $\mathbf{F} = \mathbf{dA}$, A is definitely not unique. Let

$$\mathbf{A}' = \mathbf{d}\lambda + \mathbf{A},$$

where λ is a form one lower in rank. Surely, $\mathbf{F} = \mathbf{dA}'$ also. If $\mathbf{A}$ is found by using the integral method above, the change to $\mathbf{A}'$ is obtained by a shift in the z_0^μ coordinate of the base point. The addition of $\mathbf{d}\lambda$ to $\mathbf{A}$ is called a **gauge transformation**.

Final remarks: The r-forms, because their components are antisymmetric arrays, have only $\binom{2n}{r}$ linearly independent elements at each point. Any r-form with r greater than the dimension of the space is zero. The collection of all forms of all ranks is called the **Grassman algebra**. The existence of a $2n$-form (where $2n = $ dimension of the space; remember, we are still thinking about phase space) that is everywhere nonzero is allowed only if certain global topological conditions are met. The manifold is then called **orientable**, and with a little thought you can see how this existence of an everywhere nonzero form of highest rank does fit a commonsense definition of orientable. (We will tacitly assume that phase space is orientable.)

EXERCISES

9.1. (a) Prove that $du/dt = [u, H] + \partial u/\partial t$ by writing down the Hamilton equations and the definition of the Poisson bracket. (Here u is a function of the coordinates q^i, the momentum components p_i, and t.)

(b) If u and v are constants of the motion, so is $[u, v]$. Prove this fact, allowing for the possibility that both $\partial u/\partial t$ and $\partial v/\partial t$ may be nonzero. (*Hint:* You may find Jacobi's identity useful.)

9.2. Given the Hamiltonian $H = q^1 p_1 - q^2 p_2 - a(q^1)^2 + b(q^2)^2$, where a and b are constants, show that

$$F_1 \equiv \frac{p_2 - bq^2}{q^1}, \qquad F_2 = q^1 q^2, \qquad F_3 \equiv q^1 e^{-t}$$

are constants of the motion. Discuss their independence and whether any other independent constants of the motion exist. If such exist, exhibit some until you have the maximum number of independent ones. Show explicitly that the Poisson brackets $[F_i, F_j]$ are constants of the motion.

9.3. Differential forms in $\mathbf{R}^3$:

(a) Let $f(x, y, z) = x^2 + y^2 - z$. Find $\mathbf{d}f$.

(b) Let $\boldsymbol{\omega} = x^2 \, \mathbf{d}x$. Prove that $\mathbf{d}\boldsymbol{\omega} = 0$.

(c) Let $\boldsymbol{\theta} = z \, \mathbf{d}x \wedge \mathbf{d}y$. Find $\mathbf{d}\boldsymbol{\theta}$.

(d) Find $\boldsymbol{\theta} \wedge \mathbf{d}f$.

(e) Prove that $\mathbf{d}\boldsymbol{\alpha} = 0$, where $\boldsymbol{\alpha} = y \, \mathbf{d}x \wedge \mathbf{d}z + x \, \mathbf{d}y \wedge \mathbf{d}z$.

(f) In what sense are there forms $\boldsymbol{\mu}, \boldsymbol{\nu}$ such that $\boldsymbol{\omega} = \mathbf{d}\boldsymbol{\mu}, \boldsymbol{\alpha} = \mathbf{d}\boldsymbol{\nu}$? Find $\boldsymbol{\mu}$ and $\boldsymbol{\nu}$.

9.4. (a) In $\mathbf{R}^3$ with rectangular coordinates, let f be a function. $\vec{\nabla} f$ is a vector field. Is it a covariant or a contravariant vector field? (*Hint:* Prove that it is covariant.)

(b) Let $\vec{A}$ be a vector field, considered as a covariant vector field or differential form. Compute $\mathbf{d}\vec{A}$, a 2-form. Also compute $\vec{\nabla} \times \vec{A}$. Show how $\vec{\nabla} \times \vec{A}$ and $\mathbf{d}\vec{A}$ are related.

(c) An antisymmetric tensor $\mathbf{T}$ of second covariant rank in $\mathbf{R}^3$ has three independent components. It corresponds to a 2-form. Let $B_x = T_{yz}, B_y = T_{zx}, B_z = T_{xy}$. Compute $\mathbf{d}\mathbf{T}$ and show the relation to $\vec{\nabla} \cdot \vec{B}$.

10

CANONICAL TRANSFORMATIONS

Canonical transformations are a particular class of coordinate transformations in phase space that maintain the form of Hamilton's equations, making only additive changes to the Hamiltonian H. It is fairly simple to make coordinate transformations on phase space that destroy the form of Hamilton's equations. Coordinate transformations in phase space that give $\bar{p}, \bar{x}$ that obey Hamilton's equations for a perhaps new (but only additively so in the sense explained below) Hamiltonian $\bar{H}(\bar{p}, \bar{x})$ [here $H(\bar{x}, \bar{p}) = H\big(x(\bar{x}), p(\bar{p})\big)$] are called **canonical transformations**.

Let $\{q^i, p_i, t\}$ be the coordinates of phase spacetime $\mathcal{P} \times \mathbf{R}$. (We return later to the notation z^μ for the coordinates.) We presume the existence of a Hamiltonian $H(q, p, t)$ and of the Hamilton equations

$$\dot{q}^i = \frac{\partial H}{\partial p_i}, \qquad \dot{p}_i = -\frac{\partial H}{\partial q^i}, \qquad \dot{t} = 1, \qquad \dot{H} = \frac{\partial H}{\partial t}. \qquad (10.1)$$

We now suppose a change of coordinates (typically, this change works only over a limited region of $\mathcal{P}$, but we suppress for now a discussion of this point). The new coordinates will be called Q^i, P_i, T:

$$Q^i = Q^i(q, p, t), \qquad P_i = P_i(q, p, t), \qquad T = T(q, p, t). \tag{10.2}$$

The transformation is canonical if there exists a function $K(Q, P, T)$ so that the actual paths of the system are given by

$$\dot{Q}^i = \frac{\partial K}{\partial P_i}, \qquad \dot{P}_i = -\frac{\partial K}{\partial Q^i}, \qquad \dot{T} = 1, \qquad \dot{K} = \frac{\partial K}{\partial t}, \tag{10.3}$$

where K is additively related to H [meaning that $K = H +$ something; Eq. (10.9) makes this precise].

A transformation that works in this way for an H but not for all H's will be called **canonical with respect to** $\boldsymbol{H}$. A truly canonical transformation will be one that preserves the form of Hamilton's equations no matter what specific H is used.

First, the requirement $\dot{t} = 1$ originally let us identify t (as a spacetime coordinate) with the path parameter. We do the same with T for the same reason, so that instead of T depending on q^i, p_i, we have simply

$$T = t. \tag{10.4}$$

Next, we recall that the actual paths in phase spacetime are those that extremize the path integral of the Lagrangian.

$$\delta \int (p_i \dot{q}^i - H) \, dt = 0. \tag{10.5}$$

Such paths also extremize the Lagrangian formed from K:

$$\delta \int (P_i \dot{Q}^i - K) \, dt = 0 \tag{10.6}$$

if the transformation is to be canonical. The integrand of (10.5) or of (10.6) is simply a particular "Lagrangian," grist for the mill of the equivalent Lagrangians of Chapter 5. The integrand of Eq. (10.5), for instance, depends on $(q, \dot{q}, p, t)$; it is thus distinguished in that it is only first degree (in fact, linear) in $\dot{q}$ and does not depend on $\dot{p}$ at all. (Here we are considering p_i and q^i completely independent variables.) In multidimensional systems, there is no adequate theory of equivalent Lagrangians. We do know, however, that extremum paths for the integral of $p_i \dot{q}^i - H$ and the extremum paths for the integral of $P_i \dot{Q}^i - K$ will coincide (for general H) when the integrands differ by a nonzero multiplicative factor A and an additive total time derivative dF/dt.

$$p_i \dot{q}^i - H = A\left(P_i \dot{Q}^i - K + \frac{dF}{dt} \right). \tag{10.7}$$

This relation is reasonable, at least, when we look at the path integral for any path

$$\int (p_i \dot{q}^i - H) \, dt = A \int (P_i \dot{Q}^i - K) \, dt + F(P_b) - F(P_a), \tag{10.8}$$

where P_a, P_b are the (fixed) endpoints of the curve. Clearly, a given path is an extremum path for the left integral if and only if it is an extremum path for the right integral, if the relation between integrands is as given.

The special case $A = 1$ is sufficient to show the structure of canonical transformations. It is this choice that makes K additively related to H. E. J. Saletan and A. H.Cromer (*Theoretical Mechanics,* Wiley, New York, 1971) call such a transformation a *restricted transformation*, but other authors define canonical transformations to obey $A = 1$. For this section we assume that $A = 1$ and find sufficient richness with this restriction:

$$p_i \dot{q}^i - H = P_i \dot{Q}^i - K + \frac{dF}{dt}. \tag{10.9}$$

The function F is called the **generating function**. It may be expressed as a function of any set of independent variables. However, some very convenient results are obtained if F is written as a function of n old variables and n new variables, plus time. The results are especially convenient if the n old variables include either all n of the q^i or all n of the p_i, and if the new variables include either all n of the Q^i or all n of the P_i.

For example, F may be thought of as a function of Q^i, P_i, t. Because the set $\{P, Q, t\}$ and the set $\{p, q, t\}$ both span $\mathcal{P}$ and because there are relations connecting the two sets, mixed sets such as (Q^i, q^i, t) are generically independent coordinates on $\mathcal{P}$, and we may express F as $F = F(Q, q, t)$.

The phase spacetime coordinates $\{Q^i, q^i, t\}$ are not a canonical set but are used strictly for convenience. This form of the generating function is called the F_1 form by Goldstein (*Classical Mechanics,* 2nd ed., Addison Wesley, Reading, Massachusetts, 1980):

$$F_1 = F_1(Q, q, t). \tag{10.10}$$

It is not always the case that $\{q, Q\}$ are chosen as the independent pair. It may be that other mixtures of old and new variables are needed or are convenient. The other functional forms are

$$F_2 = F_2(P, q, t), \tag{10.11}$$

$$F_3 = F_3(Q, p, t), \tag{10.12}$$

$$F_4 = F_4(P, p, t). \tag{10.13}$$

Thus suppose that $F = F_1(Q, q, t)$. We multiply the relation (10.9) between integrands by $\mathbf{d}t$ to obtain

$$P_i \, \mathbf{d}Q^i - K \, \mathbf{d}t + \left(\frac{\partial F_1}{\partial t}\right) \mathbf{d}t + \left(\frac{\partial F_1}{\partial Q^i}\right) \mathbf{d}Q^i + \left(\frac{\partial F_1}{\partial q^i}\right) \mathbf{d}q^i = p_i \, \mathbf{d}q^i - H \, \mathbf{d}t. \tag{10.14}$$

Our assumptions in this case are equivalent to the statement that the 1-forms $\mathbf{d}Q^i, \mathbf{d}q^i, \mathbf{d}t$ are independent. Hence we find that

$$P_i = -\frac{\partial F_1}{\partial Q^i}, \qquad p_i = \frac{\partial F_1}{\partial q^i}, \qquad K = H + \frac{\partial F_1}{\partial t}. \tag{10.15}$$

F_1 therefore contains an arbitrary additive constant, for its derivatives affect the transformation.

The equation in (10.15) for p_i, for example, may be inverted to give Q^i as a function of q, p, t. Then the equation for P_i becomes expressible in terms of q, p, t. Since K must be expressed in terms of Q, P, t, the variables in H and $\partial F_1/\partial t$ must be changed accordingly.

We now investigate the other forms, Eqs. (10.11)–(10.13), for F. To obtain an expression for which $F_2(P, q, t)$ is useful, we must rewrite the basic expression relating the integrands in a form that involves $\mathbf{d}P_i$ and $\mathbf{d}q^i$. The necessary step is to write, first,

$$P_i \, \mathbf{d}Q^i = \mathbf{d}(P_i Q^i) - Q^i \, \mathbf{d}P_i. \tag{10.16}$$

Then the function $P_i Q^i$ is added into the definition of F_2; we take

$$F_2 = F_1 + P_i Q^i$$

and

$$-Q^i \, \mathbf{d}P_i - K \, \mathbf{d}t + \frac{\partial F_2}{\partial t} \, \mathbf{d}t + \frac{\partial F_2}{\partial P_i} \, \mathbf{d}P_i + \frac{\partial F_2}{\partial q^i} \, \mathbf{d}q^i = p_i \, \mathbf{d}q^i - H \, \mathbf{d}t. \tag{10.17}$$

This expression is appropriate when q^i, P_i, t are independent, so that

$$Q^i = \frac{\partial F_2}{\partial P_i}, \qquad p_i = \frac{\partial F_2}{\partial q^i}, \qquad K = H + \frac{\partial F_2}{\partial t}. \tag{10.18}$$

Similarly, the other two cases result in

$$P_i = -\frac{\partial F_3}{\partial Q^i}, \qquad q^i = -\frac{\partial F_3}{\partial p_i}, \qquad K = H + \frac{\partial F_3}{\partial t}, \tag{10.19}$$

and

$$Q^i = \frac{\partial F_4}{\partial P_i}, \qquad q^i = -\frac{\partial F_4}{\partial p_i}, \qquad K = H + \frac{\partial F_4}{\partial t}. \tag{10.20}$$

The reader should check these results and should show that F_3 and F_4 may each be obtained from F_1 by adding a suitable function. Clearly, the usefulness of expressing the generating function as a function of old and new variables stems from the fact the $P_i \dot{Q}^i$ and $p_i \dot{q}^i$ appear on opposite sides of the basic relation.

Notice that the transition between different types of generating functions involves a step such as adding $-\mathbf{d}(P_i Q^i)$ to $P_i \, \mathbf{d}Q^i$ to form $-Q^i \, \mathbf{d}P_i$. A hint of what we will do in Chapter 11 may be had by taking the curl of both sides of

$$P_i \, \mathbf{d}Q^i = \mathbf{d}(P_i Q^i) - Q^i \, \mathbf{d}P_i$$

to find that

$$\mathbf{d}P \wedge \mathbf{d}Q^i = -\mathbf{d}Q^i \wedge \mathbf{d}P_i,$$

which of course is true: Both $P_i \, \mathbf{d}Q^i$ and $-Q^i \, \mathbf{d}P_i$ have $\mathbf{d}P_i \wedge \mathbf{d}Q^i$ as curl. Now take the curl of both sides of the basic relation written in the form

$$p_i \, \mathbf{d}q^i - H \, \mathbf{d}t = P_i \, \mathbf{d}Q^i - K \, \mathbf{d}t + \mathbf{d}F. \tag{10.21}$$

The result is

$$\mathbf{d}p_i \wedge \mathbf{d}q^i - \mathbf{d}H \wedge \mathbf{d}t = \mathbf{d}P_i \wedge \mathbf{d}Q^i - \mathbf{d}K \wedge \mathbf{d}t, \tag{10.22}$$

since $\mathbf{dd}F = 0$. However, the equations

$$\frac{\mathbf{d}H}{dt} = \frac{\partial H}{\partial t} \qquad \text{and} \qquad \frac{\mathbf{d}K}{dt} = \frac{\partial K}{dt} \tag{10.23}$$

mean that

$$\mathbf{d}H = \frac{\partial H}{\partial t} \, \mathbf{d}t \qquad \text{and} \qquad \mathbf{d}K = \frac{\partial K}{\partial t} \, \mathbf{d}t. \tag{10.24}$$

Consequently, $\mathbf{d}H \wedge \mathbf{d}t$ is proportional to $\mathbf{d}t \wedge \mathbf{d}t$, which vanishes, and also $\mathbf{d}K \wedge \mathbf{d}t = 0$. We therefore find that

$$\mathbf{d}p_i \wedge \mathbf{d}q^i = \mathbf{d}P_i \wedge \mathbf{d}Q^i. \tag{10.25}$$

This equality of 2-forms will play a central role in what follows. The 2-form

$$\Omega = \mathbf{d}p_i \wedge \mathbf{d}q^i \tag{10.26}$$

is called the **symplectic 2-form**. Equation (9.5) gives a coordinate expression for this object. (Note the implied sum over i.)

Some simple examples will illustrate the procedure. Consider as generating function

$$F_2 = q^i P_i.$$

We find that

$$Q^i = q^i, \qquad p_i = P_i, \qquad K = H,$$

so that this F_2 generates the identity transformation. As an exercise find a function of the form $F_3 = F_3(Q, p, t)$ that also generates the identity transformation, showing that two apparently different functions can generate the same canonical transformation.

Now let

$$Q^i = Q^i(q), \qquad \text{with} \qquad \det \left| \frac{\partial Q^i}{\partial q^j} \right| \neq 0,$$

so that these functions define a legitimate coordinate transformation in configuration space (time independent for convenience). We define an F_2 generating function by

$$F_2 = Q^i(q) P_i.$$

The transformation generated is

$$Q^i = Q^i(q), \qquad p_i = P_s \frac{\partial Q^s}{\partial q^i}, \qquad K = H.$$

The equation for p_i is the same as the equation for the transformation of the components of a 1-form. We therefore have

$$p_i \, \mathbf{d}q^i = P_i \, \mathbf{d}Q^i,$$

which is somewhat stronger than the statement of (10.25).

We can see more directly that an arbitrary change of configuration space coordinates can be used to define a canonical transformation. We use $Q^i = Q^i(q)$ as the coordinate transformation. In this case we have

$$\dot{Q}^i = \frac{\partial Q^i}{\partial q^s} \dot{q}^s. \tag{10.27}$$

We then define $\bar{L}(Q, \dot{Q}, t)$ by the numerical equivalence

$$\bar{L}(Q, \dot{Q}, t) = L(q, \dot{q}, t) \tag{10.28}$$

(the functional form of $\bar{L}$ will in general be different from that of L). Since $Q(q)$ can be interpreted as a simple renaming of points in configuration space, this numerical equality ensures that an actual path (an extremizing path for $\int L \, dt$) is mapped to another actual path (an extremizing path for $\int \bar{L} \, dt$).

More explicitly we consider the Euler-Lagrange equations for L:

$$\frac{d}{dt}\left(\frac{\partial L}{\partial \dot{q}^i}\right) - \frac{\partial L}{\partial q^i} = 0. \tag{10.29}$$

First, we find that

$$\frac{\partial L}{\partial q^i} = \frac{\partial \bar{L}}{\partial q^i} = \frac{\partial \bar{L}}{\partial Q^s}\frac{\partial Q^s}{\partial q^i} + \frac{\partial \bar{L}}{\partial \dot{Q}^s}\frac{\partial \dot{Q}^s}{\partial q^i}, \tag{10.30}$$

where

$$\frac{\partial \dot{Q}^s}{\partial q^i} = \frac{\partial^2 Q^s}{\partial q^i \, \partial q^t} \dot{q}^t. \tag{10.31}$$

Also,

$$\frac{\partial L}{\partial \dot{q}^i} = \frac{\partial L}{\partial \dot{Q}^s}\frac{\partial \dot{Q}^s}{\partial \dot{q}^i} = \frac{\partial L}{\partial \dot{Q}^s}\frac{\partial Q^s}{\partial \dot{q}^i}. \tag{10.32}$$

We use the fact that

$$\frac{d}{dt}\frac{\partial Q^s}{\partial q^i} = \frac{\partial^2 Q^s}{\partial q^i \, \partial q^t} \dot{q}^t \tag{10.33}$$

to find

$$\frac{d}{dt}\frac{\partial L}{\partial \dot{q}^i} - \frac{\partial L}{\partial q^i} = \frac{\partial Q^s}{\partial q^i}\left(\frac{d}{dt}\frac{\partial \bar{L}}{\partial \dot{Q}^s} - \frac{\partial \bar{L}}{\partial Q^s}\right). \tag{10.34}$$

Since $\partial Q^s/\partial q^i$ is an invertible matrix, the Euler-Lagrange equations for L are equivalent to the Euler-Lagrange equations for $\bar{L}$, precisely by a coordinate transformation. The numerical equality $\bar{L} = L$ is the same as $K = H$, showing that Hamilton's equations not only are preserved (because the Euler-Lagrange equations are) but retain numerical equality for the Hamiltonian.

So we see that arbitrary configuration space coordinate changes are (or can be made) canonical. One reason is that in phase space a canonical transformation, which transforms actual paths into actual paths, includes the prescription that allowed paths be transformed into allowed paths. The configuration space transformation automatically maps allowed paths into allowed paths because of the equation

$$\dot{Q}^i = \frac{\partial Q^i}{\partial q^s}\, \dot{q}^s,$$

which implies that

$$p_i = P_s \frac{\partial Q^s}{\partial q^i}.$$

In contrast, many phase space coordinate transformations are not canonical.

Another simple example: Consider the transformation

$$Q^i = p_i, \qquad P_i = -q^i.$$

Then

$$\mathbf{d}Q^i \wedge \mathbf{d}P_i = -\mathbf{d}p_i \wedge \mathbf{d}q^i = \mathbf{d}q^i \wedge \mathbf{d}p_i.$$

This transformation is canonical and interchanges the role of coordinates and momenta: We leave as an exercise the finding of the generating function.

A final remark: If the phase space is coordinated by $y^1 = q^1$, $y^2 = p_1$, $y^3 = q^2$, $y^4 = p_2 \ldots$, then $\Omega = \omega_{\alpha\beta}\, dy^\alpha \wedge dy^\beta$ and

$$\omega_{\alpha\beta} = \begin{bmatrix} 0 & 1 & & & \\ -1 & 0 & & & \\ & & 0 & 1 & \\ & & -1 & 0 & \\ & & & & \ddots \end{bmatrix} \tag{10.35}$$

[see Eq. (9.5)].

If one considers general coordinate transformations on the z^α, one can in general obtain components for Ω that do not have this simple form. However, if the coordinates are canonical, $\omega_{\alpha\beta}$ will have the special form given by (10.35).

Consider a two-dimensional configuration space. Then phase space is four-dimensional and

$$\Omega = \mathbf{d}p_1 \wedge \mathbf{d}q^1 + \mathbf{d}p_2 \wedge \mathbf{d}q^2. \tag{10.36}$$

This represents a surface element (see Chapter 9) that has projection onto only some of the coordinate 2-planes. The fundamental property of a canonical transformation is that it leaves Ω invariant. Consider now the algebraic process of multiplying Ω by itself. For instance, we can construct the 4-form $\Omega \wedge \Omega$:

$$\Omega \wedge \Omega = 2\, \mathbf{d}p_1 \wedge \mathbf{d}q^1 \wedge \mathbf{d}p_2 \wedge \mathbf{d}q^2. \tag{10.37}$$

(It is straightforward to see, as mentioned in Chapter 9, that in an n-dimensional phase space, the highest rank of a nonzero form is n; hence there are no other forms that can be made from Ω in a four-dimensional phase space.) Since $\Omega \wedge \Omega$ is constructed from Ω, an invariant under canonical transformations, $\Omega \wedge \Omega$ is itself invariant under such transformations. In general, one may write such products with up to n factors of Ω (if the phase space is $2n$-dimensional). The structure, which has n factors, is proportional to the volume element, analogous to that of (9.27). We thus have that the volume element is invariant under canonical transformations. A common notation is

$$\Omega^{\wedge p} \equiv \Omega \wedge \cdots \wedge \Omega \qquad [\, p \text{ factors}].$$

The objects $\Omega^{\wedge p}$ are called **Poincaré invariants**, and $\Omega^{\wedge n}$ is the *volume element*.

EXERCISES

10.1. A Hamiltonian for a simple system is $H(p, q) = Ap + Bq^2$, A and B being constants.

(a) Is there a Lagrangian for this system? If so, give it.

(b) The Hamilton equations are easily solved for this Hamiltonian. Obtain the behavior of $p(t)$ and $q(t)$. If you obtained a Lagrangian in part (a), show that the result is equivalent to the result from Lagrange's equations.

(c) A straightforward canonical transformation takes this system to a more conventional Hamiltonian, which is quadratic in the new momenta. Give the explicit form of the generating function and of the canonical transformation from $\{p, q\}$ to $\{P, Q\}$. Again, discuss the Lagrangian as in part (a).

(d) Redo part (b) for this new Hamiltonian. If you obtained a Lagrangian in part (c), show the equivalent results via Lagrange's equation. Compare your results to those of part (b) and show their equivalence (or lack of equivalence).

10.2. Let $\mathcal{C}$ be a closed path in the phase space of a given system and let

$$I = \int_{\mathcal{C}} p_i \, dx^i.$$

Prove that under a canonical transformation I remains invariant.

10.3. The figure shows two masses m_1, m_2, which lie on a frictionless surface. The two masses are connected by a spring (with spring constant k_2) and m_1 is connected to a fixed (immovable) wall by a spring (spring constant k_1).

(a) What is the dimension of the configuration space $\mathcal{M}$? Discuss the topology of $\mathcal{M}$ (e.g., is it a line, a plane, a sphere, or part of one of these?) What is the dimension of the phase space, $\mathcal{P}$?

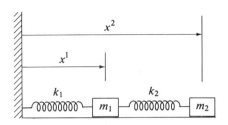

Figure P10.3

(b) Write down the Lagrangian and the Euler-Lagrange equations for this system.

(c) Obtain an explicit representation of the canonical momenta of this system and an explicit expression for the Hamiltonian, H. Is H conserved?

(d) What are the physical interpretations of the coordinates and momenta that appear in H? Exhibit a generating function and perform a canonical transformation to center of mass coordinates: That is, let one new coordinate be the position of the center of mass of the system and let the other new coordinate be the distance between the two masses. Obtain the explicit form of the Hamilton equations in terms of the new variables.

10.4. The nonrelativistic form of the Lagrangian describing a particle moving in a electromagnetic field is

$$L_{NR} = \tfrac{1}{2}m\dot{x}^i\dot{x}^i - q\phi + q\dot{x}^i A_i \qquad (i = 1, 2, 3).$$

(a) Demonstrate explicitly and precisely how this form may be obtained from the relativistically correct Lagrangian. Give reasons for each step involved.

(b) Recalling the relevant formula, $\mathbf{B} = \nabla \times \mathbf{A}$, find a form for $\mathbf{A}$ that gives rise to a uniform, constant $\mathbf{B}$ field in the z-direction.

(c) Write the nonrelativistic Lagrangian describing a three-dimensional isotropic oscillator with spring constant k (use cartesian coordinates).

(d) Now write the Lagrangian for this oscillator if the mass point has charge q and there is a uniform $\mathbf{B}$ field in the z-direction. Write down the Hamilton equations for this system.

(e) Find a solution for this set of equations. (Assume an $e^{i\omega t}$ dependence; you may also find it easier to work with the second-order equations obtained by eliminating the momenta.) What is the natural frequency of oscillations in the $(x-y)$-plane? What is the natural frequency of oscillations in the z-direction?

(f) Perform a canonical transformation with a generator of the form

$$F = (x^1 \cos \Omega t + x^2 \sin \Omega t)P_1 + (-x^1 \sin \Omega t + x^2 \cos \Omega t)P_2 + x^3 P_3$$

on this system, where $\Omega = \pm qB/2m$. What are the equations of motion in terms of the new variables, and what oscillation frequencies are obtained in this case?

10.5. (a) What is the change with time in the volume, the volume in momentum space, and the volume in phase space that is occupied by a group of equal mass particles moving freely along the x-axis? At $t = 0$ the particle coordinates are uniformly distributed in the interval $x_0 < x < x_0 + \Delta x$, and their momenta are uniformly distributed in the range $p_0 < p < p_0 + \Delta p$. Plot the region in phase space occupied by the particles for $t = 0$, $t = t_0 > 0$ (a moderate time in the future) and for a large time in the future.

   (b) Do the same for particles that move along the x-axis between two walls. Collisions with the walls are absolutely elastic. The particles do not interact with one another.

   (c) Do the same for a group of identical harmonic oscillators.

10.6. Consider a system that consists of a solid body in 3-space with one point fixed. Find the canonical transformation corresponding to new axes rotating about the z-axis with an arbitrary time-dependent angular velocity; use cylindrical coordinates.

10.7. Consider a Hamiltonian system with the Hamiltonian $H = \sqrt{p^2/2m + kx^2/2}$. Obtain the Hamiltonian equations of motion and write down their solution. Discuss (briefly) the behavior of the solution. Draw a phase space diagram for this system and indicate the solution in the diagram.

11

INFINITESIMAL CANONICAL TRANSFORMATIONS

11.1 The Hamilton–Jacobi Theory

A specific example of a generating function F is

$$F_2 = P_i\, q^i, \qquad (11.1)$$

which gives

$$p_i = \frac{\partial F_2}{\partial q^i} = P_i, \qquad (11.2a)$$

$$Q^i = \frac{\partial F_2}{\partial P_i} = q^i; \qquad (11.2b)$$

and

$$K = H. \qquad (11.2c)$$

This function thus generates the identity transformation.

We now want to consider generating functions that are "close" to F_2; we will consider only first-order contributions in some small parameter, say, $\varepsilon \ll 1$. Take a generating function

$$F_2 = P_i\, q^i + \varepsilon G(P, q, t); \tag{11.3}$$

here we assume that G has uniformly bounded first derivatives. Then

$$p_i = P_i + \varepsilon\, \frac{\partial G}{\partial q^i}, \tag{11.4}$$

$$Q^i = q^i + \varepsilon\, \frac{\partial G}{\partial P_i}. \tag{11.5}$$

These equations show that the new variables P, Q differ from the old by terms that are infinitesimal (first order in ε). Because of the assumed smoothness of G, we can write G as a function of the original phase spacetime coordinates:

$$Q^i = q^i + \varepsilon\, \frac{\partial G(p, q, t)}{\partial p_i} + \mathcal{O}(\varepsilon^2). \tag{11.6}$$

In other words, we make only higher-order errors by substituting the old variables in the argument of G. Hence, to first order,

$$P_i = p_i - \varepsilon\, \frac{\partial G}{\partial q_i} \tag{11.7}$$

$$Q^i = q^i + \varepsilon\, \frac{\partial G}{\partial p_i}. \tag{11.8}$$

We call G the **infinitesimal generating function**.

These equations have the great advantage of being explicit; that is, they do not require the elaborate functional inversion process. We also note that the changes

$$\delta p_i = P_i - p_i, \qquad \delta q^i = Q^i - q^i$$

can be summarized by

$$\begin{aligned} \delta y^\alpha &= \varepsilon \tilde{\omega}^{\alpha\beta} \partial_\beta G \\ &\equiv -\varepsilon \tilde{G}^\alpha, \end{aligned} \tag{11.9}$$

where $\tilde{\omega}^{\alpha\beta}$ is defined in Eq. (8.49); see also Eq. (9.6). We are working here in phase space (not the extended phase spacetime, which includes the time as a dynamical variable). The vector associated with an infinitesimal generating function is the vector that gives the (infinitesimal) change in coordinates under the canonical transformation resulting from the generating function.

At this point the temptation to view canonical transformations as active transformations becomes overwhelming. The infinitesimal vector field $-\varepsilon \tilde{G}^\alpha$ on phase space defines how any system moves to infinitesimally adjacent phase space locations under the transformation. Only a moment's consideration will show that Hamilton's equations in fact can be written

$$\delta y^\alpha = -\tilde{H}^\alpha \, dt. \tag{11.10}$$

That is, the Hamiltonian is the generator of **active infinitesimal canonical transformations** that take the system along its evolutionary path in time.

Suppose that the Hamiltonian is time independent. At any particular time the initial data of the system comprise a set of q_0^i, p_{0i}, namely, a point P_0 in phase space. As time evolves, the Hamiltonian carries each such system point along its particular trajectory in phase space. We can visualize the collection of all such trajectories as a **Hamiltonian flow**, and we can deduce some properties of the flow. In particular, flow lines cannot cross, and flow lines cannot join together at any point in phase space. If either of these were to happen, a particular set of initial data—the q_0^i, p_{0i} at the crossing or bifurcation point— would have nonuniquely determined future or past development.

We consider a small volume **V** defined in this flow. The phase space volume can be drawn by connecting points that are adjacent in coordinates and momenta. We know that the Hamiltonian provides a canonical transformation of the phase space as it evolves the system forward in time. We thus know that the volume just constructed will be invariant in time, and we have the ingredients to prove **Liouville's theorem**: The flow in phase space is incompressible. Our proof is simple. The volume element just defined is invariant; its boundary is defined by phase space flow lines, and no flow lines can cross the boundary or end within the volume. Hence we have a flow that is divergenceless, or incompressible.

That the Hamilton generates canonical transformations in time means specifically that the coordinates and momenta at any time are canonically transformed from the coordinates and the momenta at the initial time t_0. But the coordinates and momenta at t_0 are the initial data and are constants of the motion. They in fact give us the required number of constants of the motion to solve the dynamical problem, even if we are not able actually to carry out the solution in all cases.

The vi wpoint of the **Hamilton–Jacobi** theory is that we should find the time-dependent canonical transformation that connects the current value of the momenta and coordinates p, q to their initial values, which become the "new" coordinates and momenta, Q, P (previously called q_0, p_0). Since they are initial data, these Q, P are constants of the motion, and the dynamical equations are trivial. The difficulty has been pushed into finding the time-dependent canonical transformation.

We want a canonical transformation that follows the evolution as determined by the Hamiltonian and that transforms to the $t = t_0$ data in which the momenta and coordinates are constants. The crucial point is that the "new" coordinates and momenta are constant. We achieve this by demanding that the new Hamiltonian vanish (not just in value, but functionally):

$$K \equiv H + \frac{\partial f}{\partial t} = 0, \tag{11.11}$$

where $f(q^i, P_i, t)$ is the generating function; the P_i are the "new" momenta—that is, the constants of the motion.

If $K = 0$, we have

$$\dot{Q}^i = \frac{\partial K}{\partial P_i} = 0, \tag{11.12a}$$

$$\dot{P}_i = -\frac{\partial K}{\partial Q^i} = 0, \tag{11.12b}$$

so that the new momenta and coordinates are indeed constant. We thus seek a solution of the equation

$$0 = H(q^i, p_i, t) + \frac{\partial f(q^i, P_i, t)}{\partial t}. \tag{11.13}$$

We proceed by making use of the fact that every generating function of the form $f(q, P, t)$ generates the momentum at time t by

$$p_i = \frac{\partial f}{\partial q^i}. \tag{11.14}$$

The function f depends on the coordinates and on time; the P_i are to be constants and just act as parameters entering the equation. If we insert Eq. (11.14) into Eq. (11.13), we have the **Hamilton–Jacobi equation,**

$$0 = H\left(q^i, \frac{\partial f(q, P, t)}{\partial q^i}, t\right) + \frac{\partial f(q, P, t)}{\partial t}. \tag{11.15}$$

In the first term we mean that every appearance of p_i is replaced by $\partial f/\partial q^i$. This equation is a first-order (although in general nonlinear) partial differential equation for the generating function f. The number of independent variables is $n + 1$ (the q^i and t).

We expect an equation of this type to depend on $n + 1$ constants of integration corresponding to the initial values of the components $\partial f/\partial t, \partial f/\partial q^i$. However, we note that this equation is of a special type because only the derivatives of f (not f itself) appear in the equation. Hence if S is a solution, so is $S + \alpha$, $\alpha = $ constant. But we also know that S is a generating function and its effect is calculated only through its derivatives, so this additive constant is irrelevant. Hence S will depend on n essential constants of integration. These constants of integration are of precisely the right number and appear in the solution S just as the P_i should. The solution S is sensitive to the choice of these integration constants, and specifying these constants definitely specifies the solution. Consequently, we can identify the P_i with these constants of integration as they arise in the mathematical process of solving the differential equation. These constants will be functions of the initial data, although not generally precisely the initial momenta. (For the purposes of solving the dynamical system, the latter distinction is irrelevant.)

Once we have obtained a solution $S(q, P, t)$ of the Hamilton–Jacobi equation, then we have, from the transformation equations,

$$p_i = \frac{\partial S}{\partial q^i}, \tag{11.16}$$

which we already knew; and

$$Q^i = \frac{\partial S}{\partial P_i}. \tag{11.17}$$

The latter equation is nontrivial, since it relates Q^i to q^i, t, and the constants P_i. But the equation of motion states that Q^i is constant:

$$\dot{Q}^i = \frac{\partial K}{\partial P_i} = 0. \tag{11.18}$$

Hence the defining equation for Q^i above is, in fact, an equation that says that a certain function of q^i and of t (i.e., $\partial S/\partial P_i$) is a constant. Equation (11.17) is thus a relation between q^i and t, in other words, an explicit statement of the motion of q^i.

Consider the simple example: $H = p^2/2m$. The Hamilton–Jacobi equation is

$$\frac{1}{2m}\left(\frac{\partial f}{\partial q}\right)^2 + \frac{\partial f}{\partial t} = 0. \tag{11.19}$$

We assume that the solution is separable in the form

$$f = f_1(q) + f_2(t); \tag{11.20}$$

then

$$\frac{1}{2m}\left(\frac{df_1}{dq}\right)^2 = -\frac{df_2}{dt}, \tag{11.21}$$

which demands that each side of the equation be a constant, say, E. This is the anticipated integration constant. The solution is then

$$S(q, E, t) = \pm q\sqrt{2mE} - Et. \tag{11.22}$$

The momentum is

$$p = \frac{\partial S}{\partial q} = \pm\sqrt{2mE}. \tag{11.23}$$

Also,

$$Q = \frac{\partial S}{\partial E} = \pm\frac{qm}{\sqrt{2mE}} - t, \tag{11.24}$$

which is recognized as the statement

$$t = \frac{mq}{p} - \text{const.},\tag{11.25}$$

since $\dot{Q} = 0$ by construction of the Hamilton–Jacobi equation. Of course, (11.24) is simply the motion of a free particle.

A more interesting Hamilton–Jacobi problem is provided by the *harmonic oscillator*, based on the Hamiltonian:

$$H = \frac{1}{2m}p^2 + \frac{1}{2}m\omega^2 q^2.\tag{11.26}$$

We write the Hamilton–Jacobi equation as

$$0 = \frac{1}{2m}\left(\frac{\partial f}{\partial q}\right)^2 + \frac{1}{2}m\omega^2 q^2 + \frac{\partial f}{\partial t}.\tag{11.27}$$

Again we try a separated generating function f:

$$f = f_1(q) + f_2(t)\tag{11.28}$$

and obtain

$$-\frac{df_2}{dt} = \frac{1}{2m}\left(\frac{df_1}{dq}\right)^2 + \frac{1}{2}m\omega^2 q^2.\tag{11.29}$$

Because the left side is a function only of t while the right side is a function only of q, both sides individually equal a constant, say, $-E$. (This is the P of the generating function, arising as a separation constant in this problem.)

We have then

$$\frac{df_1}{dq} = \pm(2mE - m^2\omega^2 q^2)^{1/2},\tag{11.30}$$

which can be directly integrated. We call the solution S:

$$S = f_1(q) + f_2(t) = -Pt \pm \int^q (2mE - m^2\omega^2 q^2)^{1/2}\,dq + \alpha.\tag{11.31}$$

We have included the irrelevant additive constant α. The explicit expression for the integral

$$\int^q (2mE - m^2\omega^2 q^2)^{1/2}\,dq = \frac{1}{2}q(2mE - m^2\omega^2 q^2)^{1/2} + \frac{P}{\omega}\sin^{-1}\left(q\sqrt{\frac{m\omega^2}{2E}}\right)$$

is not needed. Notice that Eq. (11.31) gives the momentum directly:

$$p = \frac{\partial S}{\partial q} = \pm(2mE - m^2\omega^2 q^2)^{1/2}.\tag{11.32}$$

A more interesting result arises by differentiating with respect to E: the constant new coordinate Q is

$$Q = -t \pm \int^q m(2mE - m^2\omega^2 q^2)^{-1/2}\,dq$$

$$= -t \pm \frac{1}{\omega} \cos^{-1} \left(q\sqrt{\frac{m\omega^2}{2E}} \right). \tag{11.33}$$

In this equation there is still an additive constant associated with the lower limit of the integral, but we have assumed that this is included in the constant Q. We see that the coordinate Q, which is a constant of the motion, is just t_0—the time when the argument of the cosine function vanishes.

This example shows again that there are two important kinds of constants of the motion. The constant E (which we recognize as the energy) satisfies $\partial E/\partial t = dE/dt = 0$, while the constant Q (which is t_0) satisfies $dQ/dt = 0$ but $\partial Q/\partial t = -1 \neq 0$. Constants of the type of E (i.e., with no time dependence) are called **separating constants** and in general are much more useful than those of the Q type. This is because the separating constants can be separated out prior to the complete solution of the problem. The Q type can really be determined only once the total solution is known.

The Hamilton–Jacobi theory begins with Eq. (11.11) and has been outlined using a generating function $f = f(q, P, t)$, a function of type F_2 in the notation of Chapter 10. If we had chosen a function of type $F_1(q, Q, t)$, the analysis would have gone through exactly as before because P and Q are constants, and it makes no difference what we call the integration constants in the solution.

However, equations of type $F_3(Q, p, t)$ or $F_4(P, p, t)$ lead to fundamentally different Hamilton–Jacobi equations. We begin with Eq. (11.11) as before, but suppose that $f = f(P, p, t)$ with the P_i constants of the motion to be determined in the solution. Then, instead of (11.14), we use the equation valid for F_4:

$$q^i = -\frac{\partial F}{\partial p_i}. \tag{11.34}$$

The analog of Eq. (11.15) then is

$$0 = H\left(-\frac{\partial f(P, p, t)}{\partial p_i}, p_i, t \right) + \frac{\partial f(P, p, t)}{\partial t} \tag{11.35}$$

(i.e., every appearance of q_i is replaced by $-\partial f/\partial p_i$). The solution of the equation can then be carried through, the solution of the mechanical system being analogous to that given in Eq. (11.17). Since the new Hamiltonian K vanishes, we have

$$Q^i = \text{const} = \frac{\partial S(P, p, t)}{\partial P}, \tag{11.36}$$

where we again call the solution S. This gives the relationship

$$p_i = p_i(P, Q, t) \tag{11.37}$$

where Q, P are constant. Finally, we use

$$q^i = -\frac{\partial S}{\partial p_i} \tag{11.38}$$

to evaluate $q^i = q^i(P, Q, t)$.

We may, for instance, redo the two examples just given. For the free particle, Eq. (11.11) becomes

$$0 = \frac{\partial f(P, p, t)}{\partial t} + \frac{p^2}{2m}, \tag{11.39}$$

which has the immediate solution

$$S = -\frac{p^2 t}{2m} + h(p, P), \tag{11.40}$$

where h is an arbitrary function of p and P. Then we have

$$Q = \text{const} = \frac{\partial S}{\partial P} = \frac{\partial h}{\partial P}. \tag{11.41}$$

This equation can be inverted to yield

$$p = p(P, Q), \tag{11.42}$$

which simply states $p = \text{const}$. Further,

$$q = -\frac{\partial S}{\partial p} = +\frac{pt}{m} + \frac{\partial h}{\partial p} = vt + \text{const.}, \tag{11.43}$$

with $v = p/m = \text{const}$.

In the example of the free particle, the two versions of the Hamilton–Jacobi equation are qualitatively very different, although they lead to the same physical result. The harmonic oscillator, which we now treat, is essentially symmetrical between p and q, and so the development will very closely parallel that of Eqs. (11.26)–(11.33). From Eq. (11.26) we have the analog of Eq. (11.27):

$$0 = \frac{1}{2m}p^2 + \frac{1}{2}m\omega^2 \left(\frac{\partial f}{\partial p}\right)^2 + \frac{\partial f}{\partial t}, \tag{11.44}$$

where again $f = f(p, P, t)$. A separated form for f is

$$f = f_1(p) + f_2(t); \tag{11.45}$$

hence

$$-\frac{df_2}{dt} = \frac{1}{2m}p^2 + \frac{1}{2}m\omega^2 \left(\frac{df_1}{dp}\right)^2. \tag{11.46}$$

Each side of Eq. (11.46) must separately equal a constant $(-P)$; hence

$$-f_1 = \pm \int^p \left[\frac{2(P - p^2/2m)}{m\omega^2}\right]^{1/2} dp \tag{11.47}$$

and

$$S = -Pt \pm \int^p \left[\frac{2(P - p^2/2m)}{m\omega^2}\right]^{1/2} dp. \tag{11.48}$$

Equation (11.34) gives the coordinate directly:

$$q = -\frac{\partial S}{\partial p} = \mp \left[\frac{2(P - p^2/2m)}{m\omega^2}\right]^{1/2}, \tag{11.49}$$

but the interesting, dynamical equation is

$$
\begin{aligned}
Q = \text{const} &= \frac{\partial S}{\partial P} = -t \pm \int^p (m\omega^2)^{-1} \left[\frac{2(P - p^2/2m)}{m\omega^2}\right]^{-1/2} dp \\
&= -t \pm \frac{1}{\omega} \sin^{-1}\left(\frac{p}{\sqrt{2mP}}\right).
\end{aligned}
\tag{11.50}
$$

We mention here some relations that have important consequences in quantum mechanics. S is, up to simple additive terms, the total **action** of the system. For instance, consider a standard Hamilton–Jacobi solution $S(q, P, t)$ [a solution of Eq. (11.15)]. Then

$$
S(q, P, t) = \int^t L(q, \dot{q}, t') \, dt'.
\tag{11.51}
$$

Classically, this result is practically empty, since it can only be used as a sort of endpoint check on the calculations. S is a function of q, t, and the constants P, while L is a function of q, $\dot{q}$, and t, and the total action can be computed only after the path is determined, that is, only after the problem is solved.

To prove (11.51) we note that

$$
\frac{dS}{dt} = \dot{q}^i \frac{\partial S}{\partial q^i} + \frac{\partial S}{\partial t} = p_i \dot{q}^i - H = L.
\tag{11.52}
$$

The endpoint check mentioned above is to substitute the functional forms for q and $\dot{q}$ into the two sides of Eq. (11.51); if numerical agreement is not obtained, an error was made at some point in the calculation.

If S is the other kind of Hamilton–Jacobi solution, that is, if $S = S(p, P, t)$ [a solution of Eq. (11.35)], then

$$
\frac{dS}{dt} = \dot{p}_i \frac{\partial S}{\partial p_i} + \frac{\partial S}{\partial t} = -q^i \dot{p}_i - H = p_i \dot{q}^i - H - (q^i p_i)^{\cdot}.
\tag{11.53}
$$

Hence

$$
S(p, P, t) = \int_{t_0}^{t} L \, dt' - (q^i p_i)\Big|_{t_0}^{t},
\tag{11.54}
$$

as we might have anticipated from Chapter 10.

11.2 Hamilton–Jacobi Theory—Time-Independent Case

It should be apparent that for time-independent Hamiltonians, separation of variables always gives rise to an action.

$$
S = f_1(q^i) - P_t \, t;
\tag{11.55}
$$

and after the $-P_t t$ form is substituted into the Hamilton–Jacobi equation, one is left with an equation of the form

$$H\left(q, \frac{\partial f_1}{\partial q}\right) = E,$$ (11.56)

where the constant momentum P_t has been identified as the energy E, which is often the value of the Hamiltonian. An equation of the form (11.56) suggests that the generating function has transformed the system to a set of canonical variables in which the Hamiltonian is not identically zero but is equal to one of the "new" momenta. In this canonical frame,

$$\dot{P}_i = -\frac{\partial K}{\partial Q^i} = 0,$$ (11.57a)

$$\dot{Q}^t = \frac{\partial K}{\partial P_t} = 1,$$ (11.57b)

$$\dot{Q}^i = \frac{\partial K}{\partial P_i} = 0 \qquad \text{for } i \neq t.$$ (11.57c)

Notice that in this frame all of the P_i are constant, but not all the Q^i are constant; one of them, Q^t, is linear in the time.

The harmonic oscillator example proceeds almost exactly as before. We have

$$E = \frac{1}{2m}\left(\frac{df}{dq}\right)^2 + \frac{1}{2}m\omega^2 q^2,$$ (11.58)

with the solution, which we now call W (for historical reasons):

$$W(q, E) = \pm \int^q (2mE - m^2\omega^2 q^2)^{1/2}\, dq.$$ (11.59)

As before, we find that

$$p = \frac{\partial W}{\partial q} = \pm(2mE - m^2\omega^2 q^2)^{1/2}.$$ (11.60)

We also have the "new" coordinate Q^t, given by

$$Q^t = \frac{\partial W}{\partial E} = \pm \int^q m(2mE - m^2\omega^2 q^2)^{-1/2}\, dq$$
$$= \pm\frac{1}{\omega}\cos^{-1}\left(q\sqrt{\frac{m\omega^2}{2E}}\right).$$ (11.61)

The Hamilton equation (11.57b) shows, however, that Q^t is not constant, but rather $Q^t = t - t_0$, with t_0 constant. Inserting this into (11.61), we again recover the familiar harmonic oscillator solution.

In multiple-dimensional systems the W so obtained is a function of the q^i and of constants P_i. Additionally, the Hamiltonian is equal to one of the constants, or to an arbitrary linear combination of the constants. The Hamiltonian is typically equal to the total energy, which can be reexpressed in terms of some of the other constants of the motion. Hence if we view E as a function of the entire set of P_i, we may write

$$\dot{P}_i = -\frac{\partial K}{\partial Q^i} = 0; \tag{11.62}$$

$$\dot{Q}^i = \frac{\partial K}{\partial P_i} = \frac{\partial E}{\partial P_i} = \nu^i. \tag{11.63}$$

These objects ν^i are constants because they arise when E, a function of constants, is differentiated with respect to one of its arguments. The result is still a function of constants, hence it is itself a constant. We thus have

$$Q^i = \nu^i t + \beta^i = \frac{\partial W(q, P, t)}{\partial P_i}, \tag{11.64}$$

where β^i are constants. These equations give the explicit expressions for t in terms of the coordinates q^i and, when inverted, yield the q^i as functions of t.

11.3   Hamilton–Jacobi Theory and Wave Mechanics

The Hamilton–Jacobi theory is the classical expression that is the most direct limit of quantum mechanics. Consider a simple, time-independent, classical, nonrelativistic one-dimensional system as described by a Hamiltonian

$$H = \frac{1}{2m} p^2 + V(x).$$

There is a straightforward prescription for obtaining the **Schrödinger wave equation:**

$$H(x, -\hbar i \partial_x)\Psi = \left[\left(-\frac{\hbar^2}{2m} \right) \partial_x{}^2 + V(x) \right]\Psi = \hbar i \frac{\partial \Psi}{\partial t}. \tag{11.65}$$

The prescription is to replace the appearance of E in the time-independent conservation of energy equation with $\hbar i \, \partial/\partial t$ and to replace every appearance of p_i by $-\hbar i \, \partial/\partial x^i = -\hbar i \partial_x$. Here i is the imaginary number $\sqrt{-1}$, and $\hbar$ is Planck's constant, which has the units of action and is very small on a macroscopic scale ($\hbar = 1.054 \times 10^{-27}$ erg·s). The dynamical variable is the field $\Psi(x^i, t)$, which is interpreted as the probability amplitude, so that $|\Psi(x, t)|^2 \, \Delta v$ gives the probability of finding the particle in a small volume Δv centered on x at time t.

This equation is different from the Hamilton–Jacobi equation. For one thing, it is linear in Ψ; for another, it involves second derivatives of Ψ: $p^2 \to -\hbar^2 \partial_x{}^2$. In fact, there is a correspondence to classical mechanics, as we now show. By separation of variables, Eq. (11.65) has solutions Ψ, whose time dependence is $e^{-iEt/\hbar}$, where E is the energy. But because $\hbar$ is so small, the circular frequency $E/\hbar$ is, for classical systems, very large. The rate of change of Ψ in the spatial direction is also very large, since the quantity $p/\hbar$ for a macroscopic system is large.

Hence we can make the **JWKB** (for Jeffreys, Wentzel, Kramers, Brillouin) **approximation:**

$$\Psi \sim A(x)\, e^{i\phi(x,t)/\hbar}. \tag{11.66}$$

In this approximation, we assume that the phase ϕ is rapidly varying, $|\phi_{,i}|\,/\,|\phi| \gg 1$ (typically), but further derivatives of ϕ and all derivatives of $A(x)$ are small. We already know that $\phi(x,t) = \phi(x) - iEt/\hbar$. Inserting Eq. (11.66) into Eq. (11.65), we have

$$-\frac{1}{2m}\hbar^2 \left\{ -\frac{\phi_{,x}^2}{\hbar^2} + \left[\frac{2(A_{,x}/A)\,i\phi_{,x}}{\hbar} + \frac{i\phi_{,xx}}{\hbar} \right] + \frac{A_{,xx}}{A} \right\} + V = E. \tag{11.67}$$

It can be seen that under the assumptions stated for the sizes of the derivatives, the dominant terms in this equation are

$$\frac{1}{2m}\phi_{,x}^2 + V = E. \tag{11.68}$$

This is, of course, the time-independent Hamilton–Jacobi equation.

The JWKB solution can be carried to one further order. If (11.68) is satisfied, we can demand that the next order in $\hbar^{-1}$ also vanish:

$$2\left(\frac{A_{,x}}{A}\right) + \frac{\phi_{,xx}}{\phi_{,x}} = 0, \tag{11.69}$$

or

$$A^2 \propto (\phi^{-1})_{,x},$$

giving the approximate solution

$$\Psi \approx A_0\left[2m(E-V)\right]^{-1/4} e^{iEt/\hbar} \exp\left\{\frac{i}{\hbar}\int^x \left[2m(E-V)\right]^{1/2} dx\right\}. \tag{11.70}$$

We see that the Hamilton–Jacobi equation corresponds to only the most dominant leading terms in the JWKB approximation.

How does this equation in fact describe the classical motion? The Hamilton–Jacobi solution technique continues by asserting that

$$Q = \text{const} = \frac{\partial\phi(x,t)}{\partial E}, \tag{11.71}$$

and this relation between x, t, and constants defines the motion $x(t)$. Because we recognize that $\phi/\hbar$ is the phase, we recognize that the Hamilton–Jacobi equation is just the calculation of the *group velocity* of the Ψ waves. [The analogy is clouded because in the usual statement—for example, in Jackson (J. D. Jackson, *Classical Electrodynamics,* 2nd ed., Addison-Wesley, Reading, Massachusetts, 1975)—the spatial phase is some spatially constant vector times the position vector x. Here we have a generalization of that.] Hence classical particles propagate like compact packets of Ψ waves.

When further accuracy toward the quantum mechanical solution is desired, the amplitude can be approximated as in Eq. (11.70). At this level one can predict interesting phenomena, like interference between waves that have traversed different paths. Such effects are of course completely outside the realm of classical physics, but it is just such effects that led to the discovery of quantum mechanics. The pursuit of higher-order approximations to the solution of Eq. (11.70) leads to an asymptotic series, which imperfectly represents the solution. The actual quantum mechanical solution and all the wave mechanics implications are obtained from the exact solution of the wave equation, Eq. (11.65).

EXERCISES

11.1. Consider a time-dependent Hamiltonian of the form

$$H = e^t \left(\frac{1}{2m} p^2 + \frac{1}{2} m\omega^2 q^2 \right).$$

Solve the Hamilton–Jacobi equation and describe the motion.

11.2. In the presence of an electromagnetic field, the kinetic part of the Hamiltonian is $(1/2m)|\vec{p} - e\vec{A}|^2$, where $\vec{A}$ is the electromagnetic vector potential, and m and e are the mass and charge of the particle. Consider the situation in which we have a one-dimensional harmonic oscillator in a constant magnetic potential, but suppose that the zero point of the spring is moving to the right with speed $v = eA/m$, so the Hamiltonian is time dependent:

$$H = \frac{1}{2m} (p - eA)^2 + (x + vt)^2.$$

Solve the Hamilton–Jacobi problem for this system. (Try a new variable $q = x + vt$.)

11.3. Solve the Hamilton–Jacobi problem when

$$H = \frac{1}{t}(p^3 + 2px).$$

11.4. Solve the Hamilton–Jacobi problem for

$$H = (p - eA)_i \, (p - eA)^i,$$

where $A^i = \alpha x^i$, $\alpha = \text{const}$.

11.5. A Hamiltonian for a simple system is

$$H(p, q) = Ap + Bq^2,$$

where A and B are constants. Solve the Hamilton–Jacobi problem twice, once with $S = S(q, t)$, then with $S = S(p, t)$. Compare your answers to the direct solution of Hamilton's equations.

11.6. A point mass m under no external forces is attached to a weightless cord fixed to a cylinder of radius R (which is also fixed). Initially, the cord is completely wound up so that the mass touches the cylinder. A radially directed impulse is now given to the mass, which starts unwinding.

(a) Find the Lagrangian in terms of the coordinate ϕ.

(b) Set up the Hamilton–Jacobi equation $H = E$ and find the characteristic function $W(\phi, E)$.

(c) Make a canonical transformation to canonical phase space coordinates Q, E, where the new Hamiltonian is E. Solve Hamilton's equations in (Q, E)-space.

(d) Using the results of part (c), find the general solution, in terms of ϕ. Which is the solution satisfying the correct initial condition?

(e) Find the angular momentum L of the mass about the cylinder axis, using the result of part (d). Why is L not constant?

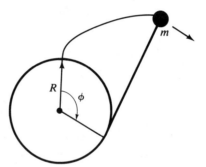

Figure P11.6

11.7. Consider the Hamiltonian

$$H = \frac{1}{2m}{p_1}^2 + \frac{1}{2m}(p_2 - kq^1)^2.$$

(a) Solve for the motion by the Hamilton–Jacobi prescription. To what physical system might this correspond?

(b) Solve this problem again by solving the canonical Hamilton's equations, and again by making the canonical transformation with $Q^1 = Ap_1$, $P_1 = B(p_1 - kq^1)$, with suitable constants A, B, solving for Q^i, P_i, and transforming back.

11.8. A single particle moves in one-dimension under the action of the potential $V = x^{2\alpha}$, where α is an integer. For large α, $v(x)$ approaches a square well. Set up the Hamilton–Jacobi equation for this problem. Describe the orbits when α is large and the energy of the particle is small. How do the orbits differ from those of a free particle bouncing between two walls?

11.9. A thin-walled pipe of radius R, mass m, is placed on an inclined plane making an angle α with the horizontal. The only external force acting on the pipe is gravitation.

(a) What is the configuration space for this system? The phase space?

(b) Write the Lagrangian for this system and obtain the Euler-Lagrange equations of motion.

(c) Obtain the Hamiltonian for this system. Is it the total energy? Write down the Hamilton equations, and show that they reduce to the Lagrange equation found in part (b).

(d) Solve the equations of motion for this system.

(e) Solve again, using the Hamilton–Jacobi approach, assuming first that $S = S(p)$ and then that $S = S(x)$.

12

SEPARABLE HAMILTONIANS AND THE ACTION–ANGLE FORMULATION

12.1 Introduction

The very simplest separable problem is that for which

$$H = \sum H_i(q^i, p_i), \tag{12.1}$$

with each summand depending only on its own coordinate and momentum. Such a system is called **completely separable** or **simply separable.** When approaching this problem from the time-independent Hamilton–Jacobi theory, we have

$$E = H\left(q^i, \frac{\partial W}{\partial q^i}\right)$$

$$= \sum_i H_i\left(q^i, \frac{\partial W}{\partial q^i}\right). \tag{12.2}$$

If we write

$$W = \sum_i W_i(q^i, P_i), \tag{12.3}$$

188

we have

$$E = \sum_i H_i\left(q^i, \frac{dW_i}{dq^i}\right). \tag{12.4}$$

The usual arguments about separability imply that the summands are separately constant:

$$H_i\left(q^i, \frac{dW_i}{dq^i}\right) = P_i \tag{12.5}$$

and

$$E = \sum_i P_i. \tag{12.6}$$

We then proceed as outlined in Chapter 11.

Suppose that we have a simply separable system that is periodic in each of its coordinates. The periods need not be identical. If they are not, the system is called **multiply periodic.** In fact, if the ratios of the frequencies are irrational, the motion will not be truly periodic—it will never return to its original position in phase space, even though each coordinate oscillates periodically.

We will now turn to a version of the Hamilton–Jacobi formulation that is adapted to the case of periodic systems. For a periodic system, the motion continually traces out the same path in phase space. The technique we now consider introduces the "new" constant momentum, here labeled by the symbol J and called an **action variable.** J is defined as

$$J_i = \oint p_i \, dq^i \qquad \text{(no sum on } i\text{)}. \tag{12.7}$$

For each coordinate there is a J_i defined. Now J_i is obviously conserved in periodic motion, since it is the area contained within the loop swept out in the q^i, p_i plane in phase space. It is further a function of the initial data for the motion. Hence it is a perfectly good "new" momentum. For completely separated systems there is in fact only one constant (e.g., E_i) associated with each coordinate. To introduce the new momentum, J_i, we must express E_i in terms of J_i:

$$E_i = E_i(J_i) = H\left(q^i, \frac{dW_i}{dq^i}\right). \tag{12.8}$$

Once this equation is solved, we then have the new cordinate Q^i defined as usual:

$$Q^i = \frac{\partial W_i}{\partial J_i}(q^i, J_i). \tag{12.9}$$

Furthermore, we know that

$$\dot{Q}^i = \frac{\partial K}{\partial J_i} = \nu^i \tag{12.10}$$

is a constant.

The Hamilton–Jacobi generating function yields the usual expression for each p_i:

$$p_i = \frac{\partial W_i(q^i, J_i)}{\partial q^i}. \tag{12.11}$$

The fundamental calculational step in this analysis is the evaluation of J_i

$$J_i = \oint p_i \, dq^i$$
$$= \oint \frac{\partial W_i}{\partial q^i} dq^i. \tag{12.12}$$

It will be noted that an indefinite integral of the form $\int (\partial W_i/\partial q^i)dq^i$ must be evaluated to determine the generating function of W_i. In (12.12) we require somewhat less: the total integral of the partial derivative around the path in phase space.

In simple systems we may proceed directly by evaluating the change in the generating function W_i as it is carried around one closed orbit. If J_i is nonzero, as we certainly expect from an examination of the phase space diagram, we are forced to conclude that W_i is multiple valued. On the other hand, W_i is a function, whereas J_i is a single number. We expect that there must be simple, shortcut methods for calculating J_i rather than going through the W_i. Since we are dealing with two-dimensional contour integrals, the theory of Cauchy integrals is appropriate. For the simple harmonic oscillator both techniques can be used and we will demonstrate them below. (As an aside, we notice that we can evaluate J_i by a straightforward examination of the simple harmonic oscillator phase space.) For more complex systems, however, only the shortcut methods will be feasible. In typical Hamiltonians in which the momentum enters quadratically, the integrand contains a radical that is real only for certain values of q^i. Physically, we know that the zeros of the radical correspond to the turning points of the motion. The momentum typically proceeds from one turning point to the other with (say) positive momentum (the positive sign of the square root), going to zero at the turning point, and the orbit is completed by motion that has negative momentum (the negative sign of the square root). The action variable J_i can thus be written

$$J_i = \int_{q^i(TP_q)}^{q^i(TP_2)} \frac{\partial W_i}{\partial q^i} dq^i. \tag{12.13}$$

For a typical motion the turning points are a function of the energy E as is the integrand. Hence we will find that J_i is a function of E, just as expected, with all the other variables integrated out.

For the simple harmonic oscillator, we have previously written the indefinite integral involved in (12.13) [cf. Eq. (11.31)]. By evaluating this integral at the turning points, we obtain

$$J = \frac{2\pi E}{\omega}. \tag{12.14}$$

Alternatively, we may evaluate the cyclic integrals as integrals over contours in a complexified x-space. The classic work in this regard is A. Sommerfeld's *Atomic Structure and Spectral Lines* (Methuen, London, 1930). The technique is discussed by Goldstein, *Classical Mechanics* (Addison-Wesley, Reading, Massachusetts, 1980). Let us evaluate the harmonic oscillator action variable in this way.

The integral to evaluate J consists of one loop from turning point q_{min} to turning point q_{max} and back again. On the first part of the integral, the square root appearing in (12.13) has positive sign, as does dq; on the second half they both have negative sign, so that the J integral is twice the integral from q_{min} to q_{max}. Note that $q_{max} > 0 > q_{min} = -q_{max}$. Alternatively, in the complex analytic treatment, we consider an integration as shown in Figure 12.1, in the complex-q plane. It is standard to consider counterclockwise contours as giving a positive integral, hence the assignment of positive sign to the square root on the lower side and negative sign on the upper. The point is that (by continuity) if the contour C closely "hugs" the real-axis interval, then

$$\oint [2mE - m^2\omega^2 q^2]^{1/2} dq \qquad (12.15)$$

obviously agrees with the basic definition of J for the harmonic oscillator. Furthermore, according to the theory of analytic functions, the contour may be distorted at will, without changing the value of the contour integral, as long as it crosses no singularities of the integrand. The region between q_{min} and q_{max} is a continuous singular interval (because the integrand has opposite sign as we approach the real axis from the two sides and thus is effectively undefined there). The region between q_{min} and q_{max} on the real axis is called a *cut*, and in fact a second copy of the complex-q plane can be attached there.

There are, happily, simpler kinds of singularities where one finds r^{-n}, $r \to 0$ behavior. This kind of singularity is called a *pole*. Now the behavior of an analytic function near a pole (at z_0) is

$$f(z) = \sum_{n=-|N|}^{\infty} a_n (z - z_0)^n, \qquad (12.16)$$

Phase is positive here

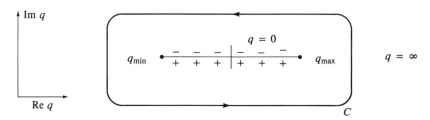

Figure 12.1

an expansion called a **Laurent series**; it is essential that $|N|$ be finite. The coefficient a_{-1} is called the **residue** of f at z_0. The point of this is that *the value of a counterclockwise contour integral of an analytic function about a pole is $2\pi i$ times the residue at the pole.* If a contour contains more than one pole, the contribution from each is counted—as if the contour were broken into several loops, one about each pole. There is no correspondingly simple integration theorem for integration along a cut, but in our case, distortion of the contour, followed by integration around the poles in the problem, will yield the answer.

The complex plane in fact has the topology of a sphere, and $r = \infty$ is an ordinary point on the sphere. [This can be seen by a "stereographic projection" from the plane onto a sphere (Figure 12.2). The sphere sits on the plane at the origin. The origin maps to the south pole. Every point on the plane is mapped by drawing a straight line from the *north* pole to the point in question on the plane. Where this line cuts the sphere is the image on the sphere of the point. The north pole is obviously the image of ∞, and the real axis is a meridian circle between the north and south poles.] The contour, which encloses the segment of the real line between $q_{\min}, q_{\max}$ obviously encloses—on its other side—all the rest of the sphere.

Now the integrand in Eq. (12.15) has residue only at ∞. (In the q-plane version, the "infinite" length of the contour contributes like a pole; more explicitly, one takes $u = 1/q$ and evaluates poles near $u = 0$.) The residue at $u = 0$ is

$$\pm i \frac{E}{\omega}. \tag{12.17}$$

To evaluate the integral, we have to carry out carefully the distortion of the contour and verify which of the signs (arising from the square root) is appropriate.

First, the distorted curve shown in Figure 12.3 becomes as shown in Figure 12.4, where in both these figures we give the "sphere picture" on the left, and the "plane picture" on the right. Notice that this last contour is traversed in a clockwise direction. The integral (12.15) is thus $(-2\pi i)$ times the residue at ∞.

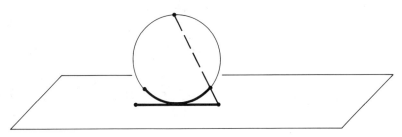

Figure 12.2

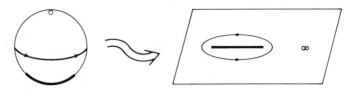

Figure 12.3

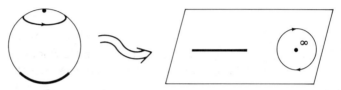

Figure 12.4

Finally, we need to know the *sign* of the square root function to evaluate correctly the residue at ∞. Return to the integral in the form (12.15). We simplify the square root by dividing by $|2mE|^{1/2}$ and consider the behavior of

$$\sqrt{1 - s^2} \equiv i\sqrt{s^2 - 1} = i\sqrt{s - 1}\,\sqrt{s + 1}, \qquad (12.18)$$

where $s^2 = m\omega^2 q^2/|2E|$.

We consider the two factors—the square roots—defined by taking their individual branch points to the left, so $\sqrt{s - 1}$ is cut from $-\infty$ to $s = 1$ while $\sqrt{s + 1}$ is cut from $s = -\infty$ to $s = -1$. (Our analysis closely follows that of P. M. Morse and H. Feshbach, *Methods of Theoretical Physics,* McGraw-Hill, New York, 1953, Section 4.4.)

For this choice of definition of the factors, Figure 12.5 then gives a direct geometrical meaning to the factors. Now for a point just above the real axis between $s = -1, s = +1$ we have

$$\psi_- = \text{phase of } \sqrt{s - 1} \simeq \frac{i\pi}{2},$$

$$\psi_+ = \text{phase of } \sqrt{s + 1} \simeq 0.$$

Hence above the real axis in the interval $(-1, 1)$, the phase of the product $\sqrt{s^2 - 1}$ is $\psi = \psi_- + \psi_+ = i\pi/2$, so $\sqrt{s^2 - 1} \simeq i|s^2 - 1|^{1/2}$.

Just below the real axis, in the interval $(-1, 1)$,

$$\psi_- \simeq -\frac{i\pi}{2},$$

$$\psi_+ \simeq 0,$$

$$\psi \simeq -\frac{i\pi}{2}$$

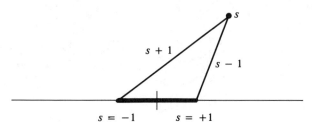

Figure 12.5

(i.e., $\sqrt{s^2 - 1} \simeq -i|s^2 - 1|^{1/2}$). On the real axis, for $s > 1$, the phase of each of the factors, and of the product, is zero. Just above the real axis, for $s < -1$,

$$\psi_- \simeq \frac{i\pi}{2},$$

$$\psi_+ \simeq \frac{i\pi}{2},$$

$$\psi \simeq i\pi$$

and $\sqrt{s^2 - 1} \simeq -|s^2 - 1|^{1/2}$. Just below the real axis, for $s < -1$,

$$\psi_- \simeq -\frac{i\pi}{2},$$

$$\psi_+ \simeq -\frac{i\pi}{2},$$

$$\psi \simeq -i\pi,$$

which is the same result as before and $\sqrt{s^2 - 1} = -|s^2 - 1|^{1/2}$. Hence the cuts in the factors "cancel out" on the real axis $s < -1$.

Multiplying back the i factored out in Eq. (12.18), we have

$$\sqrt{1 - s^2} \simeq -|1 - s^2|^{1/2}$$

above the real axis, in the interval $(-1, 1)$;

$$\sqrt{1 - s^2} \simeq +|1 - s^2|^{1/2}$$

below the real axis, in the interval $(-1, 1)$;

$$\sqrt{1 - s^2} = i|1 - s^2|^{1/2}$$

on the real axis $s > 1$; and

$$\sqrt{1 - s^2} = -i|1 - s^2|^{1/2}$$

on the real axis $s < -1$. Exactly the same results are obtained, of course, if the factor i is not orginally factored out of Eq. (12.18), (cf. Exercise 12.7). The approach given here was used because it allows a simpler diagram.

We have just found that the sign of the square root near ∞ is given by $+i|1 - s^2|^{1/2}$. When we evaluate the residue by changing variables to $u = 1/q$, we introduce one last minus sign: $du = -dq/q^2$. The residue at $+\infty$ is thus

$$\frac{iE}{\omega}. \tag{12.19}$$

Hence

$$J_i = \frac{2\pi E}{\omega}. \tag{12.20}$$

This is identical to that obtained by a direct integration of the phase space.

The utility of the Hamilton–Jacobi approach can be seen by the following exercise. The coordinate conjugate to the momentum J_i is historically denoted w^i, and given the name **angle variable.** Now

$$w^i = \frac{\partial W_i}{\partial J_i}(q^i, J_i) \tag{12.21}$$

and

$$\dot{w}^i = \frac{\partial K}{\partial J_i}$$

$$= \frac{\partial E(J_i)}{\partial J_i} \tag{12.22}$$

$$= \nu^i(J_i)$$

with ν^i constant. Since the system is periodic, we may follow the evolution of w^i as q^i moves around its closed orbit. We have, from (12.21),

$$\Delta w^i = \oint \frac{\partial}{\partial q^i} \frac{\partial W_i}{\partial J_i} \, dq^i \qquad \text{no sum on } i \tag{12.23}$$

as the change in w^i as q^i moves through one complete period. Interchanging differentiations yields

$$\Delta w^i = \frac{\partial}{\partial J_i} \oint \frac{\partial W_i}{\partial q_i} \, dq^i$$

$$= \frac{\partial}{\partial J_i} J_i \tag{12.24}$$

$$= 1.$$

Hence we have that w^i [which, according to Eq. (12.22), is linear in time: $w^i = \nu^i t + \beta^i$, β^i constant] satisfies:

$$
\begin{aligned}
w^i(t + \tau^i) &\equiv (t + \tau^i)\nu^i + \beta^i \\
&= w^i(t) + 1 \\
&\equiv \nu^i t + \beta^i + 1,
\end{aligned}
\tag{12.25}
$$

where τ^i is the period of the ith separated part of the system. Algebra yields

$$
\nu^i \tau^i = 1.
\tag{12.26}
$$

This is the fundamental, central result of the action-angle theory. The constants ν^i are the fundamental frequencies of the motion. In many situations one may be interested in the periods of the motion without desiring a full solution to the problem. The action-angle technique shows how to carry out this program. We need only evaluate E in terms of J_i to carry out the procedure.

The action-angle formulation is a version of the time-independent Hamilton–Jacobi formalism in which the momenta are defined to be particular functions of the constants of the motion. We assume that the Hamiltonian under consideration is separable and periodic in each variable. We have already noted that each J_i is equal to the area of phase space contained in the closed path in the two-dimensional phase space associated with the ith conjugate variables. Now

$$
\begin{aligned}
\oint_{\partial R} p_i \, \mathbf{d}q^i &= \int_R \mathbf{d}(p_i \, \mathbf{d}q^i) = \int_R \mathbf{d}p_i \wedge \mathbf{d}q^i = \int_R \mathbf{d}P_i \wedge \mathbf{d}Q^i \\
&= \oint_{\partial R} P_i \, \mathbf{d}Q^i,
\end{aligned}
\tag{12.27}
$$

where the penultimate step assumes that P_i, Q^i are canonical variables, as are p_i, q^i. Hence the action variable can be written in terms of any canonical set that maintains the original separated form of the problem.

The path in phase space clearly depends on the initial constants, and the area contained within the phase space path (i.e., J_i) is certainly a function of the constants of the motion. We mentioned above that the mathematical formulation usually suggests convenient forms for the integration constants that arise in the solution of the Hamilton–Jacobi equation. The action-angle formalism requires that these be reexamined in terms of the J_i. The Hamilton equation for w^i is [cf. Eq. (12.10)]

$$
\begin{aligned}
\dot{w}^i &= \frac{\partial K_i}{\partial J^i} \\
&= \nu^i \quad (\text{const.}).
\end{aligned}
\tag{12.28}
$$

We have already seen that ν^i is $(\tau^i)^{-1}$, where τ^i is the period of the ith coordinate motion. Consequently, if one can calculate J_i and express the energy in terms of J_i, the period of the oscillation can be calculated by differentiation.

Another important property of the J_i is that they are **adiabatic invariants.** Suppose that $H = H(q, p, a(t))$, where $\dot{a}(t)/a$ is much less than τ^{-1} for τ any period in the system. Then it turns out that $\dot{J} \propto \dot{a}^2$, so that J is constant if a change occurs adiabatically (i.e., arbitrarily slowly).

An example of an adiabatic change is given by a pendulum with a string of variable length. We know from common experience that the frequency of the pendulum will increase if the string is shortened. The process of shortening the string changes the period and also does work on the string.

The tension T in the string (defined to be positive) has two terms: the term that corresponds to the weight of the mass on the string, which gives a term $mg \cos \theta$ (along the string), and a second term that arises from the centrifugal force: $mv^2/\ell = m\ell \dot{\theta}^2$. Now the work done in shortening the string by an amount $\Delta \ell$ is

$$\Delta E = -\int_{\ell}^{\ell + \Delta \ell} |T|\, d\ell \tag{12.29}$$

and if the shortening is adiabatic, we can make the approximation

$$\Delta E \simeq -\langle |T| \rangle \Delta \ell, \tag{12.30}$$

where $\langle |T| \rangle$ is the average of the tension over many oscillations of the pendulum and this is work done on the pendulum if $\Delta \ell < 0$. If we assume small oscillations, the pendulum behaves like a harmonic oscillator, and we find that

$$\langle T \rangle = \langle mg \cos \theta + m\ell \dot{\theta}^2 \rangle$$

$$\simeq mg \left\langle 1 - \frac{\theta^2}{2} \right\rangle + m\ell \langle \dot{\theta}^2 \rangle \tag{12.31}$$

$$\approx mg - \frac{mg\theta_{\max}^2}{4} + \frac{m\ell \dot{\theta}_{\max}^2}{2} + O(\theta^3).$$

The square of the maximum angle, $\theta_{\max}^2$, and the square of the maximum angular velocity, $\dot{\theta}_{\max}^2$, are related:

$$\dot{\theta}_{\max}^2 = \omega^2 \theta_{\max}^2$$

$$= \frac{g}{\ell}\, \theta_{\max}^2. \tag{12.32}$$

Hence

$$\Delta E \approx -\left(mg + \frac{mg\theta_{\max}^2}{4} \right) \Delta \ell. \tag{12.33}$$

The first term clearly represents the work involved in lifting the pendulum bob through the distance $\Delta \ell$. The other term represents a transfer of energy to the internal motion of the pendulum [note that the energy of oscillations is $\frac{1}{2}m(\ell^2 \dot{\theta})_{\max}^2 = \frac{1}{2}mg\ell\, \theta_{\max}^2$]:

$$\Delta E = -\Delta \ell \frac{E}{2\ell}. \tag{12.34}$$

Hence

$$\frac{\Delta E}{E} = -\frac{1}{2}\frac{\Delta \ell}{\ell}$$

$$= \frac{\Delta \nu}{\nu},$$

(12.35)

or

$$\frac{E}{\nu} = \text{const}.$$

(12.36)

Now it is straightforward [and we have calculated in Eq. (12.20)] that the action variable for a harmonic oscillator is

$$J = \frac{E}{\nu}.$$

(12.37)

The important point is that the action variable is an adiabatic invariant.

A second example system consists of a point particle undergoing one-dimensional motion between two perfectly elastic but infinitely massive walls. If the walls are fixed, the particle executes one full cycle in a time

$$\tau = \frac{2\ell}{v}$$

(12.38)

(where v is the speed of the particle). If the walls move inward, the particle energy changes slightly because every wall encounter adds to its energy. The wall encounters are elastic; if the wall is moving inward with speed $V \ll v$, then

$$E\Big|_{\text{AFTER}} = \frac{m}{2}(v + 2V)^2$$

$$\simeq E\Big|_{\text{BEFORE}} + 2mv\,V.$$

(12.39)

Hence in one cycle the change in kinetic energy (after two bounces of the walls during which time the change in length between the walls is $\Delta \ell$) is

$$\Delta E \simeq -4mv\,V$$

$$\simeq -2mv\,\frac{\Delta \ell}{\tau}$$

$$\simeq -mv^2\,\frac{\Delta \ell}{\ell}.$$

(12.40)

Hence

$$\frac{\Delta E}{2E} \simeq -\frac{\Delta \ell}{\ell},$$

(12.41)

where the sign reflects the increase in energy as the walls move inward. In other words,

$$\ell E^{1/2} \simeq \text{const}.;$$

(12.42)

$\ell E^{1/2}$ is an adiabatic invariant.

The action variable J is easy to calculate for this case:

$$
\begin{aligned}
J &= \oint p \, dx \\
&= 2mv\,\ell \\
&= 2^{1/2}m^{1/2}\ell E^{1/2}.
\end{aligned}
\tag{12.43}
$$

Again, the action variable is an adiabiatic invariant.

12.2   Proof of the Adiabatic Invariance of Action Variables

We have a Hamiltonian,

$$
H = H\big(q, p, a(t)\big).
\tag{12.44}
$$

We suppose that we can find a generating function

$$
S = S\big(q, J, a(t)\big),
\tag{12.45}
$$

where if a is constant, S takes us to action-angle variables:

$$
p = \frac{\partial S}{\partial q},
\tag{12.46}
$$

$$
w = \frac{\partial S}{\partial J},
\tag{12.47}
$$

$$
J = \oint p \, dq,
\tag{12.48}
$$

$$
\dot{w} = \nu.
\tag{12.49}
$$

Recall [Eq. (12.12)] that J is the increase in S over one cycle. All these statements above are true if $a = \text{const}$. If $\dot{a} \neq 0$, they are only approximately true. Now introduce a new canonical transformation given by

$$
\begin{aligned}
S^* &= S - wJ \\
&= S^*(x, w, a).
\end{aligned}
\tag{12.50}
$$

S^* differs from S by a Legendre transformation and is thus not a function of J but of w. The utility of S^* is the following: Suppose that $a = \text{const}$. Then

$$
J = \text{const}.
\tag{12.51}
$$

Now

$$
\Delta S^* \Big|_{\text{one cycle}} = \Delta S \Big|_{\text{one cycle}} - J\Delta w \Big|_{\text{one cycle}}.
\tag{12.52}
$$

Now we know $\Delta w = 1$ over one cycle, and J is equal to the change of S in one cycle. Hence

$$\Delta S^* \Big|_{\text{one cycle}} = 0. \tag{12.53}$$

If $\dot{a} \neq 0$, this result is only approximate. In that case, we can calculate

$$J = -\frac{\partial S^*}{\partial w}$$

$$\dot{J} = -\frac{d}{dt} \frac{\partial S^*}{\partial w}$$

$$= -\frac{\partial}{\partial w} \frac{dS^*}{dt} \tag{12.54}$$

$$= -\left(\frac{\partial}{\partial w} \frac{\partial S^*}{\partial a} \right) \dot{a}.$$

Integrate this equation over time:

$$\Delta J = -\int_0^\tau \dot{a} \frac{\partial}{\partial w} \frac{\partial S^*}{\partial a} \, dt. \tag{12.55}$$

Under the assumption that $\dot{a}$ changes slowly compared to the periodicities of the system, we find that

$$\frac{\Delta J}{\dot{a}} = -\int_0^\tau \frac{\partial}{\partial w} \frac{\partial S^*}{\partial a} \, dt. \tag{12.56}$$

The proof proceeds by showing that the right side is proportional to $\dot{a}$; hence ΔJ is proportional to $\dot{a}^2$, which is the statement that J changes little when a changes slowly.

The function S^* is periodic in t and, since $w = \nu t + \beta$, periodic in w with a period that changes slowly as a changes. Hence the derivative $\partial_w \partial_a S^*$ is periodic in t. Hence the integrand can be written as a Fourier series.

$$\frac{\Delta J}{\dot{a}} = -\sum_{\mu \neq 0} \int_0^\tau f_\mu(J, a) e^{2\pi i w \mu} \, dt. \tag{12.57}$$

The lack of the $\mu = 0$ term is a reflection of the fact that a w derivative has been taken. Now $w = \nu t + \beta$, but we must allow that all variables may depend on a:

$$w = \nu(a)t + \beta(a). \tag{12.58}$$

If $a = \text{constant}$, the integral over many cycles (divided by the time of integration) will vanish. If $\dot{a} \neq 0$, we insert the lowest order behavior in $\dot{a}$ (assume that $a = 0$ at $t = 0$)

$$\frac{\Delta J}{\dot{a}} = -\sum_{\mu \neq 0} \int \left[f_\mu(J, 0) + \dot{a} t \partial_a f_\mu(J, 0) \right] \tag{12.59}$$

$$\times \exp\left[2\pi i \mu (\nu_0 t + \beta_0 + \partial_a \nu_0 \dot{a} t^2 + \partial_a \beta \dot{a} t) \right] dt.$$

The terms that do not contain $\dot{a}$ average out because of the periodicity of the integral. There then remain terms of order $\dot{a}$ and higher order in $\dot{a}$. The terms involving $\dot{a}$ are proportional to $\int \dot{a} t^n [F] \, dt$, where $[F]$ is a function that is bounded and oscillatory. These integrals are small and are of order $\dot{a}$. Hence

$$\Delta J \underset{\sim}{\propto} \dot{a}^2 B, \tag{12.60}$$

where B are bounded terms, and if the change is done arbitrarily slowly, $\dot{a} \to 0$, adiabatic invariants do not change.

The royal road to quantization in the Bohr quantum theory was to equate any (action variable) adiabatic invariant to nh, where n is an integer and h is Planck's constant.

12.3 Example: The Kepler Problem as an Action-Angle Problem

For the central inverse-square force law:

$$H = \frac{1}{2m}\left(p_r^2 + \frac{p_\theta^2}{r^2} + \frac{p_\phi^2}{r^2 \sin^2 \phi}\right) - \frac{k}{r}. \tag{12.61}$$

We introduce the generating function $W(r, \theta, \phi; J_r, J_\theta, J_\phi)$. We write the Hamilton–Jacobi equation:

$$E = \frac{1}{2m}\left[\left(\frac{\partial W}{\partial r}\right)^2 + \left(\frac{\partial W}{r\partial \theta}\right)^2 + \left(\frac{\partial W}{r \sin\theta \partial\phi}\right)^2\right] - \frac{k}{r}. \tag{12.62}$$

The problem can be solved by separation of variables if we write

$$W = W_r(r) + W_\theta(\theta) + W_\phi(\phi). \tag{12.63}$$

Multiplying the Hamilton–Jacobi equation by $r^2 \sin^2 \theta$ separates the term $(\partial W/\partial f)^2 \equiv (dW_\phi/d\phi)^2$, which then equals a constant, ℓ_ϕ^2, which is obviously the square of the z-component of angular momentum. Inserting this into the Hamilton–Jacobi equation and multiplying by $2mr^2$ gives

$$2mr^2\left(E - \frac{k}{r}\right) - r^2\left(\frac{dW_r}{dr}\right)^2 = \frac{\ell_\phi^2}{\sin^2 \theta} + \left(\frac{dW_\theta}{d\theta}\right)^2. \tag{12.64}$$

The left side is a function only of r, the right side a function only of θ. Hence they must be separately equal to a constant, ℓ^2 (the square of the total angular momentum). We thus have

$$W = W_r(r) + W_\theta(\theta) + \ell_\phi \phi \tag{12.65}$$

with

$$\left(\frac{dW_\theta}{d\theta}\right)^2 + \frac{\ell_\phi^2}{\sin^2 \theta} = \ell^2 \tag{12.66}$$

and

$$
E = \frac{1}{2m} \left(\frac{dW_r}{dr} \right)^2 + \frac{\ell^2}{2mr^2} - \frac{k}{r} \tag{12.67}
$$

as the separated equations to be solved. That is, we have to integrate

$$
d\theta \left(\ell^2 - \frac{\ell_\phi^2}{\sin^2 \theta} \right)^{1/2} = dW_\theta \tag{12.68}
$$

and

$$
dr \left(2mE + 2m\frac{k}{r} - \frac{\ell^2}{r^2} \right)^{1/2} = dW_r . \tag{12.69}
$$

These equations have been reduced to direct quadratures.

The θ-integral can be written in a way that makes integration trivial (cf. Goldstein, *Classical Mechanics*, 2nd ed., Addison-Wesley, Reading, Massachusetts, 1980)

$$
W_\theta = 2\ell \int_{\pi/2-\hat\theta_0}^{\pi/2+\hat\theta_0} \left(1 - \frac{\cos^2 i}{\sin^2 \theta} \right)^{1/2} d\theta, \tag{12.70}
$$

where $\cos i = \ell_\phi / \ell > 0$ (by an adjustment of coordinates) is the cosine of the inclination of the orbit to the z-axis, and $\pi/2 \pm \hat\theta_0$ are the turning points.

Write

$$
\cos \theta = \sin i \sin \psi \tag{12.71}
$$

and

$$
u = \tan \psi \tag{12.72}
$$

(where $\psi = 0$ corresponds to $\theta = \pi/2$ and $\psi = \pi/2$ to $\theta = \pi/2 - \hat\theta_0$). Then

$$
\begin{aligned}
W_\theta &= -\ell \int_0 \frac{du}{1+u^2} \frac{\sin^2 i}{1 + u^2 \cos^2 i} \\
&= -\ell \int_0 du \left(\frac{1}{1+u^2} - \frac{\cos^2 i}{1 + u^2 \cos^2 i} \right) \\
&= -\ell \left[\psi - \cos i \arctan(\cos i \tan \psi) \right]
\end{aligned} \tag{12.73}
$$

[where ψ is the function of θ given by (12.71)].

The integration of the radial function (12.69) may also be carried out by means of elementary functions [cf. Gradshteyn and Rhzhik, *Table of Series, Integrals and Products*, ed. by A. Jeffrey, Academic Press, New York, 1980, Section (2.26)]. In principle we can use these expressions to evaluate W_r explicitly. (They are, however, extremely tedious.)

Thus we have the complete form for the generating function, which depends on the constants E, ℓ, ℓ_ϕ. The associated new coordinates

$$
Q^i = \frac{\partial W}{\partial E^i} \tag{12.74}
$$

are calculated:

$$Q^E = \frac{\partial W}{\partial E}$$

$$= t + \beta^E,$$

(12.75)

where $W = W_r + W_\theta + W_\phi$ are functions of the coordinates r, θ, ϕ and the constants E, ℓ, ℓ_ϕ. The Hamilton–Jacobi theory says that Q^E is linear in time while the other coordinates are constant. (The Hamiltonian is a function only of the single constant momentum E.)

If we temporarily give up some generality and rotate coordinates so that $\theta = \pi/2$ permanently and only ϕ changes, we have $\ell = \ell_\phi$, and

$$Q^\phi = \text{const.} = \phi - \ell_\phi \int \left[\left(E + \frac{k}{r} \right) - \frac{\ell^2}{r^2} \right]^{-1/2} \frac{dr}{r^2}.$$

(12.76)

If $u = 1/r$, the argument of the square root becomes a quadratic, and if

$$u = \frac{u_{\min} + u_{\max}}{2} + \frac{u_{\min} - u_{\max}}{2} v$$

(12.77)

(with $u_{\min \atop \max}$ determined by the turning points),

$$Q_\phi - \phi = \int \frac{dv}{(1 - v^2)^{1/2}};$$

$$v = \cos(\phi - Q^\phi).$$

(12.78)

Further, using standard definitions

$$u_{\min} = \frac{1}{a(1 + \varepsilon)}, \qquad u_{\max} = \frac{1}{a(1 - \varepsilon)},$$

(12.79)

we find that

$$u = \frac{1}{r} = \frac{1 + \varepsilon \cos(\phi - Q^\phi)}{a(1 - \varepsilon^2)}.$$

(12.80)

The evaluation of the time dependence of the motion is given via Eq. (2.49):

$$t = -\frac{m}{\ell} \int \frac{du}{u^2 \left[(u - u_{\min})(u - u_{\max}) \right]^{1/2}}.$$

This completes the solution to the problem in the Hamilton–Jacobi formalism.

The *action-angle formulation* requires that we evaluate

$$J_r = \oint \frac{dW_r}{dr} \, dr, \qquad J_\theta = \oint \frac{dW_\theta}{d\theta} \, d\theta, \qquad J_\phi = \oint \frac{dW_\phi}{d\phi} \, d\phi = 2\pi\ell_\phi.$$

(We now go back to a general situation where θ and ϕ both vary.) The two integrals J_r and J_θ can be evaluated by noting that the motion is bounded by turning points r and in θ [those values of the coordinates which make the square roots vanish in Eqs. (12.42) and (12.43)], and the cyclic integral over one orbit is equal to twice the value of the integral between the turning points. Hence we could use the Gradshteyn–Ryzhik expressions to evaluate the definite integrals:

$$J_\theta = 4\ell\left(\frac{\pi}{2}\right)(1 - \cos i)$$

$$= 2\pi(\ell - \ell_\phi), \tag{12.81}$$

$$J_r = -(J_\theta + J_\phi) - \pi k\sqrt{\frac{2m}{-E}}. \tag{12.82}$$

The more elegant technique, which works when the integral cannot be looked up in Gradshteyn and Ryzhik, is to evaluate the cyclic integrals as closed contours in a complexified r- or θ-space in the way introduced in Chapter 12 for the harmonic oscillator.

The integral to evaluate J_r consists of one loop from turning point $r_{\min}$ to turning point $r_{\max}$ and back again. On the first part of the integral, the square root appearing in (12.69) has a positive sign, as does dr; on the second half they have a negative sign, so that J_r integral is twice integral from $r_{\min}$ to $r_{\max}$. Note that both $r_{\min}$ and $r_{\max} > 0$. In the complex analytic treatment, we consider an integration as shown in Figure 12.6, in the complex-r plane. As in the harmonic oscillator case, there is a *cut* between $r_{\min}, r_{\max}$ on the real axis. We thus need to evaluate

$$J_r = \oint \frac{dr}{r}\sqrt{2mEr^2 + 2mkr - \ell^2} \tag{12.83}$$

(see Figure 12.7).

Now the integrand in Eq. (12.83) has residue at $r = 0$ and at ∞. (In the r-plane version, even though the inverse r and the $\sqrt{r^2}$ behavior approximately cancel at ∞, the "infinite" length of the contour contributes like a pole; more explicitly, one takes $u = 1/r$ and evaluates poles near $u = 0$.) The residue at $r = 0$ is (modulo signs)

$$-i\,|\,\ell\,|. \tag{12.84}$$

The residue at $r = \infty$ is

$$+ik\sqrt{\frac{m}{2|E|}}. \tag{12.85}$$

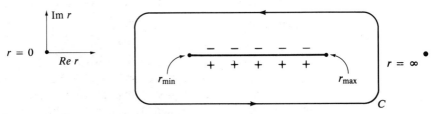

Figure 12.6

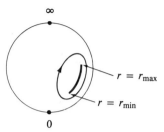

Figure 12.7

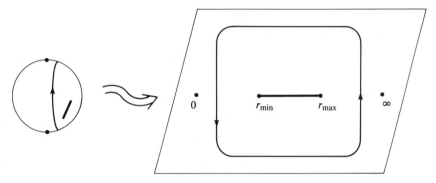

Figure 12.8

Note that $E < 0$. To evaluate the integral we have to carry out the distortion of the contour carefully and verify which of the signs (arising from the square root) is appropriate.

First the distorted curve shown in Figure 12.8 becomes as shown in Figure 12.9. The piece connecting the two small circles contains two contours very close to one another traversed in opposite directions. The integral on these pieces cancels and finally, the integral is on the contours as shown in Figure 12.10. Notice that these contours are traversed in a clockwise direction. The integral is thus $(-2\pi i)$ times the sum of the residues at 0 and ∞.

Finally, we need to know the *sign* of the square root function correctly to evaluate the residue at 0 and at ∞. The analysis follows that for the harmonic oscillator in above. In fact, because we can decide to measure a new variable, v, from an origin at $\frac{1}{2}(r_{\min} + r_{\max})$ [i.e., $v = r - \frac{1}{2}(r_{\min} - r_{\max})$], the square root has exactly the structure of the square root for the harmonic oscillator problem. Thus, on the real axis for values greater than the end of the cut, the sign of the square root is

$$i| \quad |^{1/2}, \tag{12.86}$$

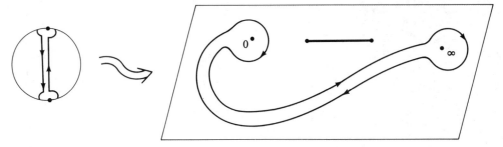

Figure 12.9

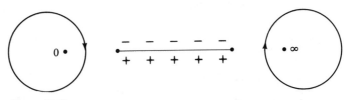

Figure 12.10

while on the real axis for values less than the left end of the cut, the sign of the square root is

$$-i| \qquad |^{1/2} . \tag{12.87}$$

Hence (12.84) as written gives the sign of the square root at $r = 0$; (12.85) gives the sign of the square root at $r = \infty$.

The residue at infinity is evaluated using the transformation $u = 1/r$, which introduces yet another sign in this term, and (12.84)–(12.87) then imply that

$$J_r = -2\pi i \left(-i\,|\,\ell\,| - ik\sqrt{\frac{m}{2|E|}} \right) \tag{12.88}$$

$$= -2\pi|\ell| - 2\pi k\sqrt{\frac{m}{2|E|}} .$$

With (12.81) we then have

$$(J_r + J_\theta + J_\phi)^2 = \frac{4m}{2|E|} k^2 \pi^2 .$$

Recognizing that $E < 0$ for bound motion, we have

$$E = -\frac{2\pi^2 k^2 m}{(J_r + J_\theta + J_\phi)^2} \tag{12.89}$$

as the expression of the Hamiltonian in terms of the constant momenta J_i. (Note that if the J_i were quantized, we would obtain the hydrogen energy spectrum $E_n \propto n^{-2}$.)

The action-angle approach yields the inverse periods of the motion exactly:

$$\nu^r = \frac{\partial E}{\partial J_r} \tag{12.90}$$

$$= \frac{4\pi^2 k^2 m^s}{(J_r + J_\theta + J_\phi)^3} \tag{12.91}$$

$$= \frac{|E|^{3/2}}{k\pi} \sqrt{\frac{2}{m}}. \tag{12.92}$$

Recalling that the semimajor axis a is related to the energy by $a = -k/(2E) > 0$, we have Kepler's law:

$$\tau_r = (\nu^r)^{-1} = 2\pi \sqrt{\frac{m}{k}} \, a^{3/2}. \tag{12.93}$$

In a multidimensional Hamiltonian system there is no guarantee that a system periodic in each coordinate will be completely (or truly) periodic. A truly periodic system must after some time (the period) return to exactly its initial state. For this to happen, the periods corresponding to each sub-Hamiltonian must be rational multiples of one another; that is, there should exist $n - 1$ sets of integers $\{j_i^\alpha\}$ (α labels the sets, i is an index that runs $1 \ldots n$) that solve

$$j_i^\alpha \nu^i = 0 \qquad \text{for each set } \alpha = 1, \ldots, k = n - 1. \tag{12.94}$$

Notice that if the system is truly periodic, there are $n - 1$ relations between the n frequencies ν^i, so there is only one independent frequency. A system that is completely periodic is also said to be completely degenerate. If only some of the periods are rationally related, the system is said to be **partially degenerate, or multiply degenerate.** This is the situation when the index α in Eq. (12.94) is bounded by a value of k positive but smaller than $n - 1$.

If the system can be separated completely [i.e., if $H = \sum_i H(x^i, p_i)$], each individual part is completely independent, and any manipulation we desire can be carried out independently on each Hamiltonian subsystem. A trivial example is the two-dimensional harmonic oscillator, whose Hamiltonian can be written

$$H = \frac{p_x^2}{2m} + \frac{m\omega_x^2}{2} x^2 + \frac{p_y^2}{2m} + \frac{m\omega_y^2}{2} y^2, \tag{12.95}$$

which obviously separates into H_x and H_y. If $\omega_x = \omega_y$, this Hamiltonian can also be expressed in plane-polar coordinates:

$$H = \frac{p_r^2}{2m} + \frac{p_\theta^2}{2mr^2} + \frac{m\omega^2}{2} r^2. \tag{12.96}$$

This Hamiltonian still allows a separable Hamilton–Jacobi solution but is not strictly separable because the variable r or p_r appears in every term. Further, the rectangular (completely separable) form still separates even with $\omega_x \neq \omega_y$. But computing H in plane-polar coordinates when $\omega_x \neq \omega_y$ yields

$$H = \frac{p_r^2}{2n} + \frac{p_\theta^2}{2mr^2} + \frac{m}{2}\, r^2(\omega_x^2 \cos^2\theta + \omega_y^2 \sin^2\theta), \qquad (12.97)$$

which does not admit a separable Hamilton–Jacobi solution.

In systems that are not strictly separable, period degeneracies can lead to trouble in the behavior of action variables. This difficulty arises in our proof of the **adiabatic invariance** of the action variable. If we consider a multidimensional system, the generating function S^* would have to be expanded in terms of all the periods of the system. If some of the periods are rationally related, there is the possibility that some terms that have net zero frequencies may survive in the sum. These will contribute secular terms to the time behavior of J. Hence in multiply periodic *degenerate* systems, the J_i are *not* automatically guaranteed adiabatic invariants.

The difficulty with the adiabatic invariants is corrected by introducing another canonical transformation from the action-angle variables, (J_i, w^j), to a new set, $(\bar{J}_i, \bar{w}^j)$. The new sets are chosen so that exactly k of the frequencies are exactly zero. The transformation will arrange that for $j = 1 \ldots k$, $\nu^j = 0$, while for $j = k+1 \ldots n$, the $\bar{w}^j$ and $\bar{J}_j$ are identical to the corresponding w^j and J_j. Hence we will be left with $n - k$ nonzero nondegenerate frequencies, together with k explicitly vanishing frequencies. The zero-frequency modes will not affect our argument about the adiabatic invariance of J_i, $i > k$, but we have no adiabatic invariance theorem for the zero-frequency modes because of their infinite period.

The procedure is to pick a generating function

$$F = F(w^j, \bar{J}_i) \qquad (12.98)$$

$$= \sum_{a=1}^{k} \bar{J}_a\, j^a{}_b\, w^b + \sum_{s=k+1}^{n} w^s\, \bar{J}_s. \qquad (12.99)$$

Then

$$\bar{w}^s = \frac{\partial F}{\partial \bar{J}_s}, \qquad (12.100a)$$

$$= j^s{}_b\, w^b, \qquad s \le k; \qquad (12.100b)$$

$$= w^s, \qquad s > k. \qquad (12.100c)$$

Now we have by explicit calculation:

$$\dot{\bar{w}}^s = j^s{}_b\, \dot{w}^b, \qquad (12.101a)$$

$$= j^s{}_b\, \nu^b = 0, \qquad s \le k; \qquad (12.101b)$$

$$\dot{\bar{w}}^s = \nu^s, \qquad\qquad s > k. \qquad\qquad (12.101c)$$

In the discussion of adiabatic invariants the zero-frequency terms are eliminated when we differentiate to obtain the action variable J. The rest of the frequencies involved are not rationally related, so no difficulties arise in carrying out the analysis.

In the Kepler problem, we saw that since the three action variables entered the Hamiltonian as the sum $J_r + J_\theta + J_\phi$, all three frequencies are the same, so

$$\nu^r - \nu^\theta = 0 \qquad\qquad (12.102)$$

and

$$\nu^r - \nu^\phi = 0. \qquad\qquad (12.103)$$

To eliminate the degeneracy in this system we can use as a generating function

$$F = \bar{J}_\theta(w^r - w^\theta) + \bar{J}_\phi(w^r - w^\phi) + \bar{J}_r \, w^r. \qquad (12.104)$$

Hence

$$\bar{w}^\theta = \frac{\partial F}{\partial \bar{J}_\theta} = w^r - w^\theta \qquad J_\theta = \frac{\partial F}{\partial w^\theta} = -\bar{J}_\theta \qquad (12.105a,b)$$

$$\bar{w}^\phi = \frac{\partial F}{\partial \bar{J}_\phi} = w^r - w^\phi \qquad J_\phi = \frac{\partial F}{\partial w^\phi} = -\bar{J}_\phi \qquad (12.106a,b)$$

$$\bar{w}^r = w^r \qquad\qquad\qquad J_r = \frac{\partial F}{\partial w^r} = \bar{J}_\theta + \bar{J}_\phi + \bar{J}_r \qquad (12.107a,b)$$

Hence the Hamiltonian is

$$H = \frac{4\pi^2 m k^2}{\bar{J}_r{}^2} \qquad\qquad (12.108)$$

and the action variables associated with the zero-frequency modes do not appear in the Hamiltonian. $\bar{J}_r$ is now a good variable to quantize because it is now an adiabatic invariant.

Note that since $\nu^r = \nu^\theta = \nu^\phi$, the motion is exactly periodic and the Kepler orbits are closed curves (ellipses); they are so simple because each cycle in θ corresponds to precisely one cycle in ϕ and in r. In the Kepler problem, the potential is precisely kr^{-1}. If the potential is not exactly proportional to r^{-1}, the frequency of motion in the r direction may not be the same as the frequency of motion in the θ and ϕ directions. (As long as the potential is spherically symmetric, the frequencies of motion in the angular directions will obey $\nu^\theta = \nu^\phi$ because the action variables corresponding to θ and ϕ both occur in the same way in H. Since $\nu^\theta = \nu^\phi$, exactly one cycle in θ corresponds to exactly one cycle in ϕ; a spatial rotation or a canonical transformation then eliminates the θ variation, say, and we see the motion is in a plane). Because $\nu^r \neq \nu^\phi$, one cycle of the angular variables does not correspond to one cycle of the r-coordinate. Hence the orbit will not come back to the same r after moving through an angle of 2π. If the frequencies are nonetheless rationally related, the orbit will eventually close. If the ν^r and ν^ϕ are not degenerate, the orbit fills all of the region in the plane between the inner and outer turning points, never returning to any point on the orbit but passing arbitrarily close to any point in the plane between the turning points in r.

Adiabatic invariants as we have developed them here have the important property that they are minimally affected when perturbations are carried out on the system. A perturbation theory, canonical perturbation theory, has been developed for periodic systems, based on a Hamiltonian formulation using the action-angle variables. Canonical perturbation theory is very similar to the Rayleigh-Schrödinger perturbation theory of quantum mechanics. This is not surprising because of the close connection implied by the old quantum theory's quantization of adiabatic invariants.

EXERCISES

12.1. Evaluation of the radial action variable J_r can be carried out explicitly if a change of variables is made (cf. M. Born, *Atomic Physics*, translated by J. Dougall, Blackie and Son Ltd., London, 1935).

Consider motion only in the plane $\theta = \pi/2$. Introduce $u = 1/r$ and rewrite the acton integral J_r to obtain

$$J_r = \oint p_r \, dr = \oint m \left(\frac{dr}{d\phi} \dot{\phi} \right) \frac{dr}{d\phi} \, d\phi$$

$$= p_\phi \int_0^{2\pi} \frac{\varepsilon^2 \sin^2 \phi}{(1 + \varepsilon \cos \phi)^2} \, d\phi, \tag{12.109}$$

where we use [Eq. (4.53)] $r = p_\phi^2/\left(km/(1 + \varepsilon \cos \phi)\right)^{-1}$. Rewrite this as

$$J_r = -p_\phi \int_0^{2\pi} \frac{\varepsilon \cos \phi \, d\phi}{1 + \varepsilon \cos \phi} \tag{12.110}$$

by integration by parts. Then evaluate $2 \times (12.109) - (12.110)$ to obtain

$$J_r = -p_\phi \cdot 2\pi + p_\phi \frac{1 - \varepsilon^2}{(p_\phi^2/km)} \int_0^{2\pi} r^2 \, d\phi. \tag{12.111}$$

The last integral is the area of the ellipse, $= \pi ab$, and the semiminor and semimajor axes can be evaluated in terms of the orbit and the ellipticity (cf. Chapter 7):

$$J_r = -2\pi p_\phi + \frac{2\pi p \phi}{1 - \varepsilon^2}.$$

Show that Eq. (12.89) follows.

12.2. Two elastic spherical particles of small radius and with masses m and M, respectively ($m \ll M$), move along the straight line OA shown in the

Figure P12.2

figure. At 0 the particle m is reflected elastically by a wall. Assume that the velocity of the lighter particle at $t = 0$ is much larger than that of the heavier particle and determine the motion of the heavier particle averaged over the period of motion of the lighter particle.

12.3. A modified Harmonic oscilator is given by

$$H = e^{-\alpha t}\left(\frac{p^2}{2m} + \frac{m\omega^2}{2}q^2\right).$$

   (a) Is the Hamiltonian the energy? Is the Hamiltonian conserved?

   (b) Write down the Hamilton's equation. A straightforward solution is obtained by considering new variables: $\hat{q} = m\omega q$; $\hat{p} = p$, and then writing an equation in terms of

$$Z = \hat{q} + i\hat{p} \qquad (i = \sqrt{-1}).$$

Give this solution.

   (c) *Solve* the Hamilton–Jacobi equation for this case. [Note that $S = S(q, P, t)$.] Solve for the motion using the generating function obtained. Show the result is the same as that for part (b).

   (d) Supposing $\alpha = 0$, what are the action-angle variables for this problem?

   (e) If α is very small, the Hamiltonian should be undergoing *adiabatic* change. Using the adiabatic invariant of the problem, can you discuss the evolution of the motion? What, specifically is the criterion of "smallness" for α?

   (f) By using the results of part (b) or (c), explicitly show that the action variable is in fact an adiabatic invariant. Can you comment on the behavior when α is not small according to the criterion in part (e)?

12.4. The Hamiltonian for a two-dimensional harmonic oscillator is

$$H = \frac{p_1{}^2}{2m_1} + \frac{p_2{}^2}{2m_2} + \omega^2[(x^1)^2 + (x^2)^2],$$

where $m_1 = m_2/2$. Write down the action-angle variables for this system. Is it degenerate? If so, transform to a system that has only one nonzero frequency. What is the meaning of the action-angle variables in this new system?

12.5. The variational principle for point particle motion in a centrally symmetric gravitational field in general relativity is

$$\delta I = 0,$$

$$I = \int \left[-f(r) \left(\frac{dt}{d\lambda} \right)^2 + \frac{1}{f(r)} \left(\frac{dr}{d\lambda} \right)^2 + r^2 \left(\frac{d\theta}{d\lambda} \right)^2 + r^2 \sin^2\theta \left(\frac{d\phi}{d\lambda} \right)^2 \right] d\lambda,$$

where

$$f(r) = 1 - \frac{2GM}{c^2 r},$$

and where λ is a parameter different from t [t is considered a separate coordinate to be solved for, but you will not need to find $t(\lambda)$ for the problem], G is Newton's constant, and c is the speed of light.

12.6. Obtain the Hamiltonian for this system and then, using the Hamilton–Jacobi formation, obtain the periods of the motion (assume that the motion is in the plane $\theta = \pi/2$). Inserting the mass of the sum, calculate the precession of the perihelion for Mercury and for the Earth.

12.7. The complex integration of expression (12.15) required that we determine the sign of the square root function at ∞, given our choice of signs on the *cut* along the real axis.

   (a) Work through the determination of the signs following Eq. (12.18), but *without* first factoring $\sqrt{-1}$ from the radical. The answer should agree with that given in the text.

   (b) From the viewpoint of complex analysis, the point ∞ is the same as both the real "points" $+\infty$ and $-\infty$ (cf. the stereographic projection). Hence we could have considered unwrapping the contour in Figure 12.1 to wrap around the value $-\infty$ at the negative end of the real line. Then Figures 12.3(b) and 12.4(b) would be different. The analysis, however, gives the same value for the integral. Demonstrate this.

12.8. A plane double pendulum is constructed as shown; each mass is the same, and the lengths of the two sections are the same. Assume that θ_1 and θ_2 remain small. Prove that there exist coordinates θ_1 and ϕ_2 that

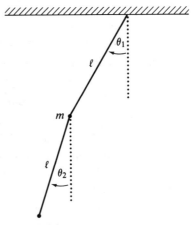

Figure P12.8

are linear combinations of θ_1 and θ_2 such that the Hamiltonian is completely separable. Then write the separated Hamilton–Jacobi equations, explicitly determining all coefficients that arise.

12.9. A particle slides on a pair of frictionless planes, which therefore make a kind of oscillator, as shown in the figure. There is no friction or other dissipation.

(a) Solve the action-angle problem for this situation and display the frequencies.

(b) Initially, the experiment has $\alpha = \alpha_0$ and $|x_{max}| = x_{max\,0}$ (the amplitude of the oscillation). What are the energy and frequency of the oscillation?

(c) Now suppose that the angle α is *slowly* changed to a new value α_1. What are the new values of the energy and the frequency?

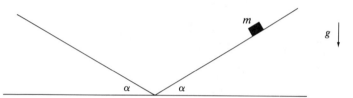

Figure P12.9

13

CANONICAL PERTURBATION THEORY

Consider a periodic system with a time-independent Hamiltonian of the form

$$H = H(q, p, \lambda),$$ (13.1)

where λ is a constant that specifies the strength of a "perturbation." We assume that

$$H_0(q, p) = H(q, p, 0)$$ (13.2)

defines a system that can be solved exactly. Note that the system is assumed to be periodic regardless of the value of λ as long as λ is in a specified interval containing zero. Because the unperturbed Hamiltonian is periodic, it may be solved (exactly) in terms of action-angle variables w^0, J_0. Thus

$$H_0(q, p) = K_0(J_0)$$ (13.3)

and

$$\dot{w}^0 = \frac{\partial K_0}{\partial J_0};$$ (13.4)

and note

$$\dot{w}^0 = \nu^0(J_0) \tag{13.5}$$

and

$$J_0 = \text{const}. \tag{13.6}$$

Now (w^0, J_0) constitute a set of canonical variables. There is thus a canonical transformation (the Hamilton-Jacobi action-angle transformation), which takes us from the variables (q, p) to (w^0, J_0). Exactly the same transformation may be written—and will still be canonical—regardless of what the Hamiltonian is. The perturbed Hamiltonian

$$H(p, q, \lambda) \tag{13.7}$$

can thus be written as

$$H(w^0, J_0, \lambda). \tag{13.8}$$

Because the perturbed Hamiltonian depends on w^0, J_0 is no longer constant if $\lambda \neq 0$. On the other hand, one can in principle obtain the action-angle variables (w, J) appropriate to the perturbed problem. Then

$$H(p, q, \lambda) = E(J, \lambda) \tag{13.9}$$

where the remaining dependence on λ arises because the appropriate action-angle variables, and hence the dependence of E on them, is different for different λ. Also,

$$\dot{w} = \frac{\partial E}{\partial J} \tag{13.10}$$
$$= \nu(J, \lambda)$$

and

$$\dot{J} = -\frac{\partial E}{\partial w} = 0.$$

Now we recall that the purpose of the Hamilton-Jacobi approach is to obtain expressions relating the variables (q, p) to (w, J), since the latter have simple time dependence. We already have the canonical transformation that connects (q, p) to (w^0, J_0); the approach we pursue here is then to find the canonical transformation (which exists by the group property of canonical transformations) that connects (w^0, J_0) to (w, J). When λ is small, we expect this transformation to be close to the identity. We will use this fact to obtain the generating function of the transformation by making an expansion in powers of λ. We thus need

$$S(w^0, J, \lambda) \tag{13.11}$$

(a different generating function for each λ). The expansion of S, which we assume to exist, gives

$$S = S_0(w^0, J) + \lambda\, S_1(w^0, J) + \lambda^2 S_2(w^0, J) + \cdots . \tag{13.12}$$

For $\lambda = 0$ we require that S be the identity, that is,

$$S_0 = w^0 J \tag{13.13}$$

[cf. Eqs. (11.1) and (11.2)]. At this point we mention certain gauge transformations that may be performed on S without affecting the canonical transformation defined by S. For instance, let

$$\tilde{S} = S + aw^0 + bJ \tag{13.14}$$

with a and b constants, independent of w^0 and J. Then

$$\begin{aligned}
w &= \frac{\partial \tilde{S}}{\partial J} \\[2mm]
&= \frac{\partial S}{\partial J} + b
\end{aligned} \tag{13.15}$$

and

$$\begin{aligned}
J_0 &= \frac{\partial \tilde{S}}{\partial w^0} \\[2mm]
&= \frac{\partial S}{\partial w^0} + a;
\end{aligned} \tag{13.16}$$

the gauge terms add only constants to the variables. Since the equation of motion for w is simply that $\dot{w} = $ constant, additive constants in w are irrelevant. Similarly, since a is a constant of integration and additive constants in J_0 are irrelevant, we henceforth will set the arbitrary constants $a = 0$, $b = 0$ if they appear in our analysis. The canonical transformations defined by S are

$$w = \frac{\partial S}{\partial J} = w^0 + \lambda \frac{\partial S_1}{\partial J}(w^0, J) + \lambda^2 \frac{\partial S_2}{\partial J}(w^0, J) + \cdots , \tag{13.17}$$

and

$$J_0 = \frac{\partial S}{\partial w^0} = J + \lambda \frac{\partial S_1}{\partial w^0}(w^0, J) + \lambda^2 \frac{\partial S_2}{\partial w_0}(w^0, J) + \cdots . \tag{13.18}$$

We pause to discuss an important property of the coordinate w^0. In the unperturbed case, since w^0 is an angle variable, we know that $\Delta w^0 = 1$ over one cycle of the periodic system. In the perturbed case, it is w that is the angle variable, so $\Delta w = 1$ over one cycle. But what is Δw^0 over one cycle in the perturbed case?

When $\lambda \neq 0$, the only thing that can be said about w^0 is that it is related to p and q by a particular canonical transformation that makes $\{w^0, J_0\}$ a canonical set. In fact, this canonical transformation would make w^0, J_0 action-angle variables when $\lambda = 0$. We can explicitly solve the transformation to get

$$w^0 = w^0(p, q). \tag{13.19}$$

This is an explicit, time-independent expression connecting w^0 to the original variables p, q. For instance, in the harmonic oscillator example that appears below, we have from (13.52)–(13.54),

$$\frac{1}{2\pi} \arctan\left(2\pi\nu_0 m\frac{q}{p}\right) = w^0. \tag{13.20}$$

Suppose now $\lambda = 0$. Given this explicit expression of w^0 in terms of q, p, we might expect that w^0 will come back to its original value when q, p return to their original values. But, because $w^0 = \nu^0 t + \beta$ when $\lambda = 0$, where ν^0 is the frequency of oscillation of the motion, what occurs in fact is that one has to choose the appropriate branch of the function $w^0(p, q)$. In the harmonic oscillator example, one is choosing the appropriate branch of the arctangent function. After one cycle, the branch chosen has jumped to give an increment of unity in w^0 (i.e., $\Delta w^0 = 1$ per cycle in the unperturbed case). Note, however, that in the *perturbed* case, we still have exactly the same expression for w^0 in terms of (p, q). The perturbed motion might have no points in common with the unperturbed phase space orbit. Still, over one cycle of the perturbed orbit, p, q will return to their starting values. Hence, up to jumps in branch, the value of w^0 at the end of the cycle will return to its initial value. Hence $\Delta w^0 (\lambda \neq 0) = N$, N being an integer that counts how many branches to jump.

Now, in the unperturbed case there are jumps, $\Delta w^0 = 1$. If the perturbation is small, we should expect the perturbed motion to be very much like the unperturbed motion. Hence, for sufficiently small $|\lambda|$,

$$\Delta w^0 = 1 \qquad \text{per cycle} \qquad (|\lambda| \text{ sufficiently small}). \tag{13.21}$$

In addition,

$$\Delta w = 1 \qquad \text{per cycle} \tag{13.22}$$

always.

Furthermore, J is an action variable. Hence

$$J = \oint p\, dq = \oint J_0\, dw^0, \tag{13.23}$$

where the integration is around the perturbed orbit and the equality of the two integrals arises because of the phase space volume-conserving property of canonical transformations. Now integrate (13.18) around one orbit:

$$\int J_0\, dw^0 = \oint J\, dw^0 + \sum \lambda^n \oint \frac{\partial S_n}{\partial w^0}\, dw^0; \tag{13.24}$$

that is,

$$J = J\,\Delta w^0 + \sum \lambda^n \oint \frac{\partial S_n}{\partial w^0}\, dw^0. \tag{13.25}$$

Since we have just seen $\Delta w^0 = 1$, this says that

$$\sum \lambda^n \oint \frac{\partial S_n}{\partial w^0}\, dw^0 = 0, \tag{13.26}$$

which we take (assume "sufficient" analyticity, as always) to mean

$$\oint \frac{\partial S_n}{\partial w^0} \, dw^0 = 0. \tag{13.27}$$

This is equivalent to saying that S_n is purely oscillatory with fundamental period unity in w^0 and with no constant term:

$$S_n = \sum_{p \neq 0} e^{ip2\pi w_0} S_{np} \tag{13.28}$$

(where the S_{np} are constant Fourier coefficients). In addition, the Hamiltonian can be expanded in λ as a function of w^0, J_0:

$$H(w^0, J_0, \lambda) = K_0(J_0) + \lambda K_1(w^0, J_0) + \lambda^2 K_2(w^0, J_0) + \cdots . \tag{13.29}$$

Since H is a known function of w^0 and J_0 for every λ, the K_i are all known. On the other hand, we have

$$\begin{aligned} H(p, q, \lambda) &= H(w, J, \lambda) \\ &= E(J, \lambda), \end{aligned} \tag{13.30}$$

which is the expression for the energy in the "new" action-angle coordinates. In this representation the Hamiltonian is independent of the coordinate w. We expand E in powers of λ:

$$E(J, \lambda) = E_0(J) + \lambda E_1(J) + \lambda^2 E_2(J) + \cdots . \tag{13.31}$$

The two equations (13.29) and (13.31) for the Hamiltonian are to be equated as power series in λ; we will then be able to give an expansion of the energies. These two expressions for the energy depend on two different sets of variables. A straightforward Taylor expansion gives

$$H(w^0, J_0, \lambda) = H(w^0, J, \lambda) + (J_0 - J)\frac{\partial H}{\partial J} + \frac{(J_0 - J)^2}{2} \frac{\partial^2 H}{\partial J^2} + \cdots . \tag{13.32}$$

Notice that we are simply making use of the *functional* form of the Hamiltonian when reexpressed in terms of the J_0, w_0 set. Every term that has a J_0 will be rewritten in terms of J by making use of the transformation (13.18) to connect the two sets of canonical variables. We will equate our explicit expression for the Hamiltonian as a function of w^0, J_0, λ from (13.32) to that for the energy as a function of J and λ. We thus obtain a set of equations that relates terms at each power of λ. Now, expanding the appearance of $J - J_0$, using the generating function, yields

$$\begin{aligned} H(w^0, J_0, \lambda) = H(w^0, J, \lambda) &+ \frac{\partial H}{\partial J}\left(\lambda \frac{\partial S_1}{\partial w^0} + \lambda^2 \frac{\partial S_2}{\partial w^0} + \cdots \right) \\ &+ \frac{1}{2}\frac{\partial^2 H}{\partial J^2}\lambda^2 \left(\frac{\partial S_1}{\partial w^0}\right)^2 + O(\lambda^3). \end{aligned} \tag{13.33}$$

Now the form $H(w^0, J, \lambda)$ is really the functional form $H(w^0, J_0, \lambda)$ evaluated at $J_0 = J$, and the partial derivatives of H in Eq. (13.33) are evaluated at $w^0, J_0 = J$. We may use Eq. (13.29) to reexpress $H(w^0, J, \lambda)$ everywhere that it appears in Eq. (13.33):

$$H(w^0, J_0, \lambda) = K_0(J) + \lambda K_1(w^0, J) + \lambda^2 K_2(w^0, J) + \cdots$$

$$+ \lambda \frac{\partial S_1}{\partial w^0} \left(\frac{\partial K_0(J)}{\partial J} + \lambda \frac{\partial K_1(w^0, J)}{\partial J} + \cdots \right) \quad (13.34a)$$

$$+ \lambda^2 \left[\frac{\partial K_0(J)}{\partial J} \frac{\partial S_2}{\partial w^0} + \frac{1}{2} \frac{\partial^2 K_0}{\partial J^2} \left(\frac{\partial S_1}{\partial w^0} \right)^2 + \cdots \right]$$

$$\equiv E(J, \lambda) \quad (13.34b)$$

$$= E_0(J) + \lambda E_1(J) + \lambda^2 E_2(J) + \cdots . \quad (13.34c)$$

Since the equations have now been expanded out as functions of J and powers of λ, we are in a position to solve for the coefficients $E_i(J)$. The expansion in terms of E_i does not involve w^0, so the appearance of w^0 in Eq. (13.34) must somehow in fact be spurious. In Eqs. (13.34) the known functions are the $K_i(w^0, J)$; the $S_i(w^0, J)$ and the $E_i(J)$ are unknown.

Equating powers of λ gives

$$E_0(J) = K_0(J) \quad (13.35a)$$

$$E_1(J) = K_1(w^0, J) + \frac{\partial S_1}{\partial w^0} \frac{\partial K_0(J)}{\partial J} \quad (13.35b)$$

$$E_2(J) = K_2(w^0, J) + \frac{\partial S_1}{\partial w^0} \frac{\partial K_1(w^0, J)}{\partial J}$$

$$+ \frac{1}{2} \left(\frac{\partial S_1}{\partial w^0} \right)^2 \frac{\partial^2 K_0(J)}{\partial J^2} + \frac{\partial S_2}{\partial w^0} \frac{\partial K_0(J)}{\partial J} . \quad (13.35c)$$

$$\vdots$$

Notice that E_0 is defined in terms of K_0. The first-order energy is given by K_1, K_0, and S_1; the second-order energy is given by K_2, K_1, K_0, and S_1 and S_2, and the structure of the equations continues to higher orders. A first glance suggests a difficulty because at first order, say, we need to know S_1 in addition to K_1 to determine E_1. The way out of this difficulty is to realize that the E_i are functions only of J (i.e., of constants of the motion). Consider, for instance, Eq. (13.35b) for the constant E_1. On the right side, the term $\partial S_1/\partial w^0$ appears multiplied by the w^0-independent term $\partial K_0/\partial J$. Hence

$$E_1 = \langle E_1 \rangle$$

$$= \langle K_1 \rangle + \frac{\partial K_0}{\partial J} \left\langle \frac{\partial S_1}{\partial w^0} \right\rangle, \quad (13.36)$$

where the averaging is over w^0 (i.e., $\langle$ any function $\rangle = \oint ($ any function $) \, dw^0$).
(We have already seen that $\langle \partial S_i / \partial w^0 \rangle = 0$ and $\Delta w^0 = 1$ over one cycle.)
Because J is constant and the S_i are periodic in w^0, we find that the J_0, although
not constant, are periodic functions of the w^0 [cf. Eq. (13.18)]. The procedure
for solving the system then becomes clear. Assume that Eqs. (13.35) have been
solved up to order $n - 1$. Averaging the nth equation, we solve for E_n. Then
we insert the now-determined value of E_n into the full equation (i.e., prior to
averaging). The only undetermined expression remaining in that equation is S_n,
and the equation can be rewritten as

$$\frac{\partial S_n}{\partial w^0} = \text{known function of } w^0 \text{ and } J. \tag{13.37}$$

We solve for the appropriate periodic S_n and then we are able to continue to
order $n + 1$, since the nth order is completely solved. Proceeding explicitly for
the first few orders, we have

$$E_0 = \langle K_0 \rangle \qquad \text{zeroth order} \tag{13.38}$$

$$\left.\begin{array}{c} E_1 = \langle K_1 \rangle \\[2mm] \dfrac{\langle K_1 \rangle - K_1}{\nu^0(J)} = \dfrac{\partial S_1}{\partial w^0} \end{array}\right\} \qquad \text{first order} \tag{13.39}$$

(where we use $\nu^0 = \partial K_0 / \partial J$).

The solution for S_1 is given by direct quadrature. At second order,

$$E_2 = \langle K_2 \rangle + \left\langle \frac{\partial S_1}{\partial w^0} \frac{\partial K_1}{\partial J} \right\rangle + \left\langle \frac{1}{2} \frac{\partial^2 K_0}{\partial J^2} \right\rangle \left(\frac{\partial S_1}{\partial w^0} \right)^2 \tag{13.40}$$

and

$$\frac{\partial S_2}{\partial w^0} = \nu_0^{-1} \left[E_2 - K_2 - \frac{\partial S_1}{\partial w^0} \frac{\partial K_1}{\partial J} - \frac{1}{2} \frac{\partial^2 K_0}{\partial J^2} \left(\frac{\partial S_1}{\partial w^0} \right)^2 \right], \tag{13.41}$$

which is again a quadrature.

The canonical perturbation theory applied to periodic classical systems
closely parallels the Rayleigh-Schrödinger perturbation scheme in wave mechanics. We notice that the energy at any particular order n is determined when
only S_{n-1} is known, and S_n can then be calculated once E_n is known. In the
wave theory, the energy is known at a particular order if the wave function is
known to one lower order. Further, the first-order energy correction is gotten by
averaging the first-order perturbations (in wave mechanics one takes the expec-
tation value of the perturbing potential using the unperturbed wave function).
Additionally, the first-order wave function is found in wave mechanics by inserting the just-found first-order energy into the perturbation scheme first-order
equation.

13.1 Canonical Perturbations: An Example

We assume that $H(p, q, \lambda)$ gives periodic motion, at least for some range in λ that includes the origin. As an example of the procedure, we consider a perturbation of the harmonic oscillator. Suppose that

$$H = \frac{1}{2m} p^2 + \frac{1}{2} (k + \lambda) q^2 \tag{13.42a}$$

$$= H_0 + \lambda H_1, \tag{13.42b}$$

where

$$H_0 = \frac{1}{2m} p^2 + \frac{1}{2} kq^2 \tag{13.43}$$

$$H_1 = \frac{1}{2} q^2. \tag{13.44}$$

Of course, we can solve the Hamiltonian exactly. But we want to check that the perturbation theory agrees with what we know the exact solution must be. So first we solve the exact problem in terms of action-angle variables:

$$\frac{1}{2m} \left(\frac{\partial W}{\partial q} \right)^2 + \frac{1}{2} (k + \lambda) q^2 = E \,; \tag{13.45}$$

hence

$$J = \oint \frac{\partial W}{\partial q} \, dq = \oint \sqrt{2mE - m(k + \lambda)q^2} \, dq = \frac{E}{\nu}, \tag{13.46}$$

where

$$\nu = (2\pi)^{-1} \sqrt{\frac{(k + \lambda)}{m}}. \tag{13.47}$$

Now

$$w = \frac{\partial}{\partial J} W(q, J) = \frac{\partial}{\partial J} \int \frac{\partial W(J, q)}{\partial q} \, dq$$

$$= \frac{\partial}{\partial J} \int \sqrt{2m\nu J - m(k + \lambda)q^2} \, dq \tag{13.48}$$

$$= m\nu \int \frac{dq}{\sqrt{2m\nu J - m(k + \lambda)q^2}}$$

$$= (2\pi)^{-1} \arcsin \left(\left[\frac{2\nu J}{k + \lambda} \right]^{-1/2} q \right). \tag{13.49}$$

In other words,

$$q = \sqrt{\frac{2\nu J}{k + \lambda}} \sin 2\pi w; \tag{13.50}$$

inserting this into $p = \partial W/\partial q$, we get

$$p = 2\pi m\nu \sqrt{\frac{2\nu J}{k + \lambda}} \cos 2\pi w$$

$$= \sqrt{2m\nu J} \cos 2\pi w. \tag{13.51}$$

For the unperturbed case ($\lambda = 0$) we have

$$E_0 = \nu_0 J_0, \tag{13.52}$$

$$q = \sqrt{\frac{2\nu_0 J_0}{k}} \sin 2\pi w^0, \tag{13.53}$$

$$p = 2\pi m\nu_0 \sqrt{\frac{2\nu_0 J_0}{k}} \cos 2\pi w^0, \tag{13.54}$$

where

$$\nu_0 = \frac{1}{(2\pi)} \sqrt{\frac{k}{m}} = \frac{1}{2\pi} \sqrt{\frac{k + \lambda}{m}} \frac{\sqrt{k}}{\sqrt{k + \lambda}}$$

$$= \nu \left(1 + \frac{\lambda}{k}\right)^{-1/2}. \tag{13.55}$$

Equations (13.50) and (13.51) are canonical transformations connecting J, w to q, p, while Eqs. (13.53) and (13.54) connect J_0, w^0 to q, p.

The procedure, according to perturbation theory, is to obtain the generating function for the canonical transformation between the variables labeled by "0" and those for nonzero λ. We have an explicit result for this transformation. By eliminating q, p between the two sets, we get J, w connected directly to J_0, w^0. From Eqs. (13.50) and (13.51) we see that

$$q^2 + \frac{p^2}{(2\pi\nu m)^2} = \frac{2\nu J}{k + \lambda}. \tag{13.56}$$

Inserting (13.53) and (13.54) into (13.56) yields

$$\frac{2\nu J}{k + \lambda} = \frac{2\nu_0 J_0}{k} \sin^2 2\pi w^0 + \frac{1}{(2\pi m\nu)^2} (2\pi m\nu^0)^2 \frac{2\nu_0 J^0}{k} \cos^2 2\pi w^0$$

$$= \frac{2\nu_0 J_0}{k} \left\{ \left(\frac{\nu^0}{\nu}\right)^2 + \sin^2 2\pi w^0 \left[1 - \left(\frac{\nu^0}{\nu}\right)^2\right] \right\} \tag{13.57}$$

$$= \frac{2\nu_0 J_0}{k} \left(1 + \frac{\lambda}{k}\right)^{-1} \left[1 + \frac{\lambda}{k} \sin^2 2\pi w^0\right].$$

Equating q/p from Eqs. (13.50) and (13.51) to q/p from Eqs. (13.53) and (13.54) gives

$$2\pi w = \arctan\left[\left(1 + \frac{\lambda}{k}\right)^{1/2} \tan 2\pi w^0\right]. \tag{13.58}$$

This is the canonical transform between J, w and J^0, w^0. To compare the results with perturbation theory, we will want to expand in powers of λ:

$$\begin{aligned} J^0 = J\Bigg[&1 + \frac{1}{2}\frac{\lambda}{k}\cos 4\pi w^0 \\ &- \frac{1}{4}\left(\frac{\lambda}{k}\right)^2\left(\cos 4\pi w^0 - \frac{1}{2}\cos 8\pi w^0\right) + \cdots\Bigg] \end{aligned} \tag{13.59}$$

$$\begin{aligned} w = w^0 &+ \frac{1}{8\pi}\frac{\lambda}{k}\sin 4\pi w^0 \\ &- \frac{1}{16\pi}\left(\frac{\lambda}{k}\right)^2\left(\sin 4\pi w^0 - \frac{1}{4}\sin 8\pi w^0\right) + \cdots. \end{aligned} \tag{13.60}$$

Also,

$$\nu = \nu^0\left[1 + \frac{1}{2}\frac{\lambda}{k} - \frac{1}{8}\left(\frac{\lambda}{k}\right)^2 + \cdots\right]. \tag{13.61}$$

Expansion of Eq. (13.58) and resummation to obtain the expression in (13.60) are the only direct techniques we are aware of to verify (13.60). This is an extremely tedious procedure. However, note that (13.57) gives $J_0 = J_0(J, w^0)$, which is, by the action-angle theory,

$$J_0 = \frac{\partial S}{\partial w^0} = J\frac{\nu}{\nu_0}\left(1 + \frac{\lambda}{k}\sin^2 2\pi w^0\right)^{-1}. \tag{13.62}$$

Integration in w^0 of the explicit expression (13.62), followed by differentiation with respect to J, gives Eq. (13.58), as it must. On the other hand, expansion of the right-hand side of (13.62) can be carried out before integration. The integrated result is then trivially differentiated with respect to J. The result is Eq. (13.60); hence (13.60) is an expansion in λ of (13.58).

Now apply canonical perturbation theory. We use the perturbed Hamiltonian

$$H = H_0 + \lambda H_1, \tag{13.63}$$

which enters the equation

$$H(w^0, J_0, \lambda) = K_0(J_0) + \lambda K_1(w^0, J_0) + \lambda^2 K_2(w^0, J_0) + \cdots. \tag{13.64}$$

Well, now,

$$H_0 = \frac{p^2}{2m} + \frac{1}{2}kq^2, \tag{13.65}$$

$$H_1 = \frac{1}{2}q^2, \tag{13.66}$$

and we have Eqs. (13.53) and (13.54) connecting (p, q) to (w^0, J_0). Hence

$$K_0 = \left(\frac{p^2}{2m} + \frac{1}{2} k^2 q^2 \right) = E_0 = J_0 \nu_0 \tag{13.67}$$

$$K_1 = \frac{1}{2} q^2 = \frac{\nu_0 J_0}{k} \sin^2 2\pi w^0 = \frac{1}{2k} \nu_0 J_0 (1 - \cos 4\pi w^0) \tag{13.68}$$

$$K_i = 0, \qquad i > 1. \tag{13.69}$$

Now

$$E_1(J) = \langle K_1 \rangle; \tag{13.70}$$

that is,

$$E_1(J) = \frac{1}{2k} \nu_0 J. \tag{13.71}$$

Also,

$$\frac{\partial S_1}{\partial w^0} = \frac{\langle K_1 \rangle - K_1}{\nu^0} = \frac{1}{2k} \frac{\nu_0 J}{\nu_0} \cos 4\pi w^0 \tag{13.72}$$

$$S_1 = \frac{J}{8\pi k} \sin 4\pi w^0. \tag{13.73}$$

We will also compute:

$$
\begin{aligned}
E_2(J) &= \langle K_2 \rangle + \left\langle \frac{\partial K_1}{\partial J} \frac{\partial S_1}{\partial w^0} \right\rangle + \frac{1}{2} \frac{\partial \nu^0}{\partial J} \left\langle \left(\frac{\partial S_1}{\partial w^0} \right)^2 \right\rangle \\
&= \langle K_2 \rangle + \frac{1}{\nu^0} \left[\left\langle \frac{\partial K_1}{\partial J} \right\rangle \langle K_1 \rangle - \left\langle \frac{\partial K_1}{\partial J} K_1 \right\rangle \right] \\
&\quad + \frac{1}{2(\nu^0)^2} \frac{\partial \nu^0}{\partial J} \left[\langle K_1^2 \rangle - \langle K_1 \rangle^2 \right].
\end{aligned}
\tag{13.74}
$$

Now

$$\frac{\partial \nu^0}{\partial J} = \frac{\partial}{\partial J} \frac{\partial}{\partial J} E_0(J) = \frac{\partial^2}{\partial J^2} (\nu_0 J) = 0. \tag{13.75}$$

Hence, with Eqs. (13.67)–(13.69)

$$
\begin{aligned}
E_2(J) &= \frac{1}{\nu_0} \left(\frac{1}{2k} \nu_0 \frac{J}{2k} \nu_0 \right) - \frac{1}{\nu_0} \frac{\nu_0}{2k} \frac{1}{2k} \nu_0 J \\
&= \frac{\nu_0 J}{4k^2} \left[1 - \left(\frac{3}{2} \right) \right] = -\frac{\nu_0 J}{8k^2}.
\end{aligned}
\tag{13.76}
$$

We thus obtain

$$E(J) = \nu_0 J + \frac{\lambda}{2k} \nu_0 J - \frac{\lambda^2 \nu_0 J}{8k^2} = \nu_0 J \left[1 + \frac{1}{2} \frac{\lambda}{k} - \frac{1}{8} \left(\frac{\lambda}{k} \right)^2 + \cdots \right]. \tag{13.77}$$

Compare the frequency this implies to Eq. (13.61). Similarly, because of Eqs. (13.74) and (13.41), we have

$$\frac{\partial S_2}{\partial w^0} = \frac{1}{(\nu_0)^2} \left(\left\langle \frac{\partial K_1}{\partial J} \right\rangle \langle K_1 \rangle - \left\langle \frac{\partial K_1}{\partial J} K_1 \right\rangle - \frac{\partial K_1}{\partial J} \langle K_1 \rangle + \frac{\partial K_1}{\partial J} K_1 \right), \quad (13.78)$$

where we use our allowed gauge freedom to subtract off the average value of $\partial S_2 / \partial w^0$. Hence

$$\frac{\partial S_2}{\partial w^{0^2}} = \frac{1}{\nu_0^2} \left[-\frac{\nu_0 2_J}{8k^2} - \frac{\nu_0}{2k} \frac{\nu_0 J}{2k} (1 - \cos 4\pi w^0) + \frac{\nu_0}{2k} \frac{\nu_0 J}{2k} (1 - \cos 4\pi w^0) \right]$$

$$= -\frac{J}{4k^2} \left(\cos 4\pi w^0 - \frac{1}{2} \cos 8\pi w^0 \right). \quad (13.79)$$

Hence we integrate

$$S_2 = -\frac{J}{16\pi k^2} \left(\sin 4\pi w^0 - \frac{1}{4} \sin 8\pi w^0 \right) \quad (13.80)$$

$$S = w^0 J + \frac{\lambda J}{8\pi k} \sin 4\pi w^0 - \frac{\lambda^2 J}{16\pi k^2} \left(\sin 4\pi w^0 - \frac{1}{4} \sin 8\pi w^0 \right), \quad (13.81)$$

whence

$$J^0 = \frac{\partial S}{\partial w^0} = J + \frac{\lambda J}{2k} \cos 4\pi w^0$$

$$- \frac{\lambda^2 J}{4k^2} \left(\cos 4\pi w^0 - \frac{1}{2} \cos 8\pi w^0 \right) + \cdots \quad (13.82)$$

$$w = \frac{\partial S}{\partial J} = w^0 + \frac{\lambda}{8\pi k} \sin 4\pi w^0$$

$$- \frac{\lambda^2}{16\pi k^2} \left(\sin 4\pi w^0 - \frac{1}{4} \sin 8\pi w^0 \right) + \cdots . \quad (13.83)$$

Compare (13.59) and (13.60).

EXERCISES

13.1. Suppose that the unperturbed form of a Hamiltonian is

$$H = H_0 = \frac{p^2}{2m} + \frac{mw^2 x^2}{2}.$$

By the method of canonical perturbations, consider the effect of adding

$$H_1 = \lambda x$$

to the Hamiltonian. Present the answers to second order. Compare your results to the exact answer.

13.2. Suppose that the Hamiltonian $H(p, q, \lambda)$ describes a perturbed Harmonic oscillator:

$$H = (\tfrac{1}{2}\, m + \lambda)p^2 + \tfrac{1}{2}\, kq^2.$$

Obtain the first-order correction to the energy, compared to that for the unperturbed problem,

$$H_0 = H(p, q, \lambda = 0)$$

$$= \frac{p^2}{2m} + \frac{1}{2}\, kq^2.$$

What is the frequency to first order?

13.3. Consider an unperturbed one-dimensional Hamilton system with

$$H_0 = \frac{p^2}{2m} + V(q)$$

with

$$V(q) = k\,|q|.$$

The motion in such a potential is periodic; note that the period is a function of energy.

(a) Write down the expression giving the period as a function of energy.

(b) Now consider a perturbed Hamiltonian and solve to second order:

$$H = H_0 + H_1$$
$$H_1 = \lambda\,|q|.$$

Compare your results to the exact solution.

13.4. Consider the same unperturbed H_0 as in (13.3). If the perturbation is now

$$H_1 = \lambda q^2,$$

solve the system to second order using canonical perturbation theory.

13.5. The unperturbed one-dimensional system consists of a point mass bouncing back and forth between perfectly reflecting walls a distance ℓ_0 apart. By canonical perturbation theory, calculate the behavior, to second order, when the walls are a distance $\ell_0 + \Delta\ell$ apart, where $|\Delta\ell/\ell_0| \ll 1$.

14

SMALL OSCILLATIONS
AND CONTINUOUS SYSTEMS

Consider a mechanical system with a finite number of degrees of freedom. Suppose that the system admits an equilibrium point. The Lagrangian (assumed time independent) for the system is taken to be $L = T(q, \dot{q}) - V(q)$, with kinetic energy quadratic in the $\dot{q}$. Suppose that $q^i = q^i_{(0)}$ is the equilibrium point. For the purposes of small deviations from equilibrium we may introduce new variables, say,

$$\eta^i = q^i - q^i_{(0)}, \tag{14.1}$$

and suppose that η^i is always "small." Then to sufficient accuracy we may take

$$T = T(q^i_{(0)}, \dot{\eta}) \tag{14.2}$$

and

$$V = V\left(q^i_{(0)} + \eta^i\right)$$

$$= V(q^i_{(0)}) + \eta^i \left.\frac{\partial V}{\partial q^i}\right|_0 + \frac{1}{2}\eta^i\eta^j \left.\frac{\partial^2 V}{\partial q^i\, \partial q^j}\right|_0 + \cdots. \tag{14.3}$$

The constant term in the expansion of the potential may of course be dropped. The term linear in η^i must vanish (i.e., $\partial V/\partial q^i|_0 = 0$) because of the definition of equilibrium. [If it did not vanish, there would be a force on the system, moving it away from $q^i_{(0)}$, contradicting the statement that $q^i_{(0)}$ is an equilibrium point.] We shall define

$$V_{ij} \equiv \frac{\partial^2 V}{\partial q^i \, \partial q^i}\bigg|_0. \tag{14.14}$$

Hence, to quadratic order in η (or $\dot{\eta}$),

$$L = \tfrac{1}{2} T_{ij}(q_{(0)})\dot{\eta}^i\dot{\eta}^j - \tfrac{1}{2}\eta^i\eta^j V_{ij}. \tag{14.5}$$

We take T_{ij} positive definite, so the kinetic energy increases with velocity. The matrix V_{ij} must also be positive definite, that is,

$$V_{ij}\xi^i\xi^j \geq 0, \tag{14.6}$$

where ξ^i are the components of an arbitrary vector and where the equality holds only if $\xi^i = 0$. If V_{ij} were not positive definite, some displacements η^i from the equilibrium would not increase the potential energy, and hence either the equilibrium is not stable or it is maintained only by the higher-order terms in the expression of V. We will assume that Eq. (14.6) holds. Note that both T_{ij} and V_{ij} are symmetric.

From Eq. (14.5) the equations of motion are

$$T_{ij}\ddot{\eta}^i + V_{ij}\eta^i = 0. \tag{14.7}$$

The form of this equation so strongly resembles that for a harmonic oscillator that we try $\eta^i = \overset{(0)}{\eta}{}^i e^{-i\omega t}$ for a so-far-undetermined ω:

$$(-\omega^2 T_{ij} + V_{ij})\overset{(0)}{\eta}{}^i = 0, \tag{14.8}$$

where $\overset{(0)}{\eta}{}^i$ are the components of a constant vector $\overset{(0)}{\boldsymbol{\eta}}$.

This is a linear homogeneous matrix equation, and the only way for it to have any solution is for the matrix multiplying $\overset{(0)}{\eta}{}^i$ to be singular (i.e., it must have the property of annulling a nonzero vector). The mathematical statement of this situation is

$$\det(-\omega^2 T_{ij} + V_{ij}) = 0. \tag{14.9}$$

Obviously, this equation will not be true for all ω; if $\omega = 0$, we have $\det V$, which does not vanish (it is positive), and if $\omega \to \infty$, we have $\sim -\omega^2 \det T$, which is negative and does not vanish because T_{ij} is positive definite. Hence it is clear that this equation determines which value of ω we should use. These arguments parallel those in Chapter 7, and eigenvalue problems are in fact frequent in physics.

Equation (14.9) is a polynomial equation in ω^2, of degree equal to n, the number of degrees of freedom of the problem. Hence there will be n

(perhaps different) values of ω^2 that solve this equation. When any one of these (say, ω_ℓ) is substituted in Eq. (14.8) we may solve for the corresponding $\overset{(0)i}{\eta}_{(\ell)}$. Because (14.8) is a linear equation, the solution clearly will determine only the direction, not the size of $|\overset{(0)}{\boldsymbol{\eta}}_{(\ell)}|$. Hence we are free to normalize the $\overset{(0)i}{\eta}_{(\ell)}$. [Only for the roots of $\omega^2_{(\ell)}$ of the polynomial equation (14.9) is the operator $-\omega^2 T_{ij} + V_{ij}$ singular; that is, only for those values of ω^2 is the set of equations (14.8) consistent.]

The vectors $\overset{(0)i}{\eta}_{(\ell)}$ satisfy certain orthogonality properties. Consider:

$$\overset{(0)i}{\eta}_{(k)}(-\omega^2_{(\ell)}T_{ij} + V_{ij})\,\overset{(0)j}{\eta}_{(\ell)} = 0. \tag{14.10}$$

Subtract

$$\overset{(0)i}{\eta}_{(\ell)}(-\omega^2_{(k)}T_{ij} + V_{ij})\,\overset{(0)j}{\eta}_{(k)} = 0. \tag{14.11}$$

Since T and V are symmetric, we find that

$$\overset{(0)i}{\eta}_{(k)}T_{ij}\,\overset{(0)j}{\eta}_{(\ell)}(\omega^2_{(k)} - \omega^2_{(\ell)}) = 0. \tag{14.12}$$

Hence if $\omega^2_{(k)} \neq \omega^2_{(\ell)}$, the constant vectors $\overset{(0)}{\boldsymbol{\eta}}_{(k)}$ and $\overset{(0)}{\boldsymbol{\eta}}_{(\ell)}$ are orthogonal under the metric $T_{ij}\,dx^i \otimes dx^j$. On the other hand, if two frequencies $\omega^2_{(\ell)}$, $\omega^2_{(k)}$ turn out to be equal, nothing is implied about the orthogonality. However, if $j \neq \ell$ and $j \neq k$ and $\omega_{(j)} \neq \omega_{(k)}$, then $\boldsymbol{\eta}_{(j)}$ is orthogonal to both $\overset{(0)}{\boldsymbol{\eta}}_{(\ell)}$ and $\overset{(0)}{\boldsymbol{\eta}}_{(k)}$. The eigenvalue-eigenvector problem then reduces to the one in the 2-space orthogonal to the $\boldsymbol{\eta}_{(j)}$, $j \neq \ell$, $j \neq k$.

Consider this as a two-dimensional degenerate problem: $\omega_1 = \omega_2$, and write $\boldsymbol{\eta}_1$ and $\boldsymbol{\eta}_2$ for the vectors in the 2-space; indices now run and sum over $1, 2$ only. Using the positivity of T_{ij} gives

$$(-\omega^2\delta_{ij} + T^{-1}_{i\ell}V_{\ell j})\,\eta_j = 0. \tag{14.13}$$

The coincidence of roots of the determinant means that

$$(\lambda_{11}\lambda_{22} - \lambda_{12}{}^2) - \omega^2(\lambda_{11} + \lambda_{22}) + \omega^4 = 0 \tag{14.14}$$

(where $\lambda_{ij} = T^{-1}_{i\ell}V_{\ell j}$) must be in the form of a perfect square. Consequently,

$$(\lambda_{11} + \lambda_{22})^2 = 4(\lambda_{11}\lambda_{22} - \lambda_{12}{}^2), \tag{14.15}$$

which can be rewritten as

$$(\lambda_{11} - \lambda_{22})^2 = -4\lambda_{12}{}^2. \tag{14.16}$$

This equation implies that

$$\lambda_{12} = 0 \tag{14.17}$$

$$\lambda_{11} = \lambda_{22}. \tag{14.18}$$

Consequently, Eq. (14.13) has the form

$$\left| \begin{pmatrix} \omega^2 & 0 \\ 0 & \omega^2 \end{pmatrix} - \begin{pmatrix} \lambda_{11} & 0 \\ 0 & \lambda_{11} \end{pmatrix} \right| \eta = 0, \tag{14.19}$$

which clearly has at least two linearly independent solutions with $\omega^2 = \lambda_{11}$, and any linear combination of these solutions is a solution. Consequently, each of the vectors corresponding to the multiple roots for ω^2 can be made orthogonal to all the others and normalized, under T_{ij}. This procedure can be extended to any number of multiple roots, using this freedom of linear combination.

We observe that for the system describing small oscillations [Eq. (14.7)], the general motion is a linear combination with arbitrary (complex) constant coefficients of the time-dependent solution vectors $\boldsymbol{\eta}_{(\ell)}$:

$$\begin{aligned} \boldsymbol{\eta} &= \sum_\ell a_\ell \, \boldsymbol{\eta}_{(\ell)} \\ &\equiv \sum_\ell a_\ell \, \overset{(0)}{\boldsymbol{\eta}}_{(\ell)} \, e^{i\omega_\ell t}. \end{aligned} \tag{14.20}$$

The a_ℓ are determined by the initial conditions:

$$\begin{aligned} \boldsymbol{\eta}\big|_{t=0} &= \sum_\ell a_\ell \, \overset{(0)}{\boldsymbol{\eta}}_{(\ell)} \\ \frac{d\boldsymbol{\eta}}{dt}\bigg|_{t=0} &= \sum_\ell i\omega_\ell \, a_\ell \, \overset{(0)}{\boldsymbol{\eta}}_{(\ell)}. \end{aligned} \tag{14.21}$$

Written out in components, there are $2n$ equations for the $2n$ constants $\operatorname{Re} a_\ell$ and $\operatorname{Im} a_\ell$.

The individual **normal modes** are purely harmonic motion with fixed frequency and often can be interpreted as particularly simple or symmetrical motion of the system. The general solution contains motion simultaneously at the different **eigenfrequencies** ω_ℓ.

14.1   Example: Coupled Harmonic Oscillators

Consider a system consisting of a mass m_1 attached by a spring of constant k to a fixed wall. To m_1 is attached a second mass m_2 through a second spring of constant k. In Figure 14.1, $x_e{}^i$ are the equilibrium lengths of the springs.

$$\mathrm{T} = \tfrac{1}{2} m_1 (\dot{x}^1)^2 + \tfrac{1}{2} m_2 (\dot{x}^1 + \dot{x}^2)^2 \tag{14.22}$$

$$V = \tfrac{1}{2} k (x^1)^2 + \tfrac{1}{2} k (x^2)^2; \tag{14.23}$$

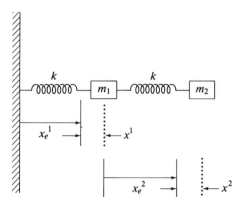

Figure 14.1

that is,

$$T = \begin{pmatrix} m_1 + m_2 & m_2 \\ m_2 & m_2 \end{pmatrix}; \qquad V = \begin{pmatrix} k & 0 \\ 0 & k \end{pmatrix}. \qquad (14.24)$$

To find the normal modes, we must solve the equation

$$(-\omega^2 T_{ij} + V_{ij}) \overset{(0)}{\eta}\,^j_{(\ell)} = 0. \qquad (14.25)$$

A solution requires that

$$0 = \det(-\omega^2 T_{ij} + V_{ij}) \qquad (14.26)$$

$$= [k - \omega^2(m_1 + m_2)](k - \omega^2 m_2) - \omega^4 m_2^2, \qquad (14.27)$$

which is, as anticipated, a quadratic equation for the frequencies of allowed motion, the eigenfrequencies. The solutions are

$$\omega_{\pm}^2 = \frac{k}{2m_2} + \frac{k}{m_1} \pm \frac{k}{2m_1 m_2}(4m_2^2 + m_1^2)^{1/2}. \qquad (14.28)$$

The eigenfrequencies are (a) a high-frequency mode (plus sign) and (b) a mode of substantially lower frequency (minus sign).

In a typical analysis we would now insert each of the frequencies ω_+, ω_- into Eq. (14.25) and solve for the direction of $\boldsymbol{\eta}_+$ and $\overset{(0)}{\boldsymbol{\eta}}_-$. The result,

$$\eta_{\pm}{}^2 = \left(\frac{k}{m_2 \omega_{\pm}^2} - \frac{m_1 + m_2}{m_2} \right) \eta_{\pm}{}^1, \qquad (14.29)$$

where $\eta_{\pm}{}^i$ are the components of $\boldsymbol{\eta}_{\pm}$. The general case (arbitrary ratio m_1/m_2) is more easily understood if we consider the limiting cases $m_1/m_2 \ll 1$ and $m_1/m_2 \gg 1$. When

$$m_1 \ll m_2;$$

$$\omega_{\pm}{}^2 \approx \frac{k}{2m_2} + \frac{k}{m_1} \pm \frac{k}{m_1} \left(1 + \frac{1}{8} \frac{m_1^2}{m_2^2} + \cdots \right). \qquad (14.30)$$

Hence

$$\omega_+{}^2 \simeq \frac{2k}{m_1} + \frac{k}{2m_2} + \frac{1}{8}\frac{k}{m_2}\left(\frac{m_1}{m_2}\right) + \cdots$$

$$\omega_-{}^2 \approx \frac{k}{2m_2} - \frac{1}{8}\frac{k}{m_2}\left(\frac{m_1}{m_2}\right) + \cdots .$$

$$(14.31)$$

The directions of the corresponding eigenvectors are given by

$$\eta_+{}^2 \simeq \left[\frac{k}{m_2}\left(\frac{m_1}{2k}\right) - \frac{m_1 + m_2}{m_2}\right]\eta_+{}^1$$

$$\approx -\eta_+{}^1$$

$$(14.32)$$

$$\eta_-{}^2 \simeq \left[\frac{k}{m_2}\left(\frac{2m_2}{k}\right) - \frac{m_1 + m_2}{m_2}\right]\eta_-{}^1$$

$$\approx \eta_-{}^1 .$$

$$(14.33)$$

Physics explains these results. Since m_1 is very small compared to m_2, if m_2 is in motion m_1 is a very small perturbation on that motion. In that case the frequency will be that of a simple harmonic oscillator of mass m and spring constant $k/2$ (i.e., $\omega = \omega_- \simeq \sqrt{k/2m_2}$). The motion of m_1 will be that of an individual point on the spring (i.e., its separation from its equilibrium position is one-half that of m_2). This is exactly what is described by $\eta_-{}^2 = \eta_-{}^1$. The other possibility in this situation is that m_2 is essentially fixed, but m_1 oscillates as if it were anchored by springs to fixed walls on both sides. The frequency then will be $\omega = \omega_+ \simeq \sqrt{2k/m_1}$. Since m_2 remains fixed, we require that $\eta_+{}^2 = -\eta_+{}^1$.

A similar analysis holds for $m_2 \ll m_1$. The two modes in this case are: m_1 oscillates and m_2 rides along without compressing its spring. The frequency is then $\omega = \omega_- \simeq \sqrt{k/m_1}$, and we find that $\eta_-{}^2 \ll \eta_-{}^1$. The second mode has m_1 essentially motionless, $\omega = \omega_+ \approx \sqrt{k/m_2}$, and $\eta_+{}^1 \ll \eta_+{}^2$.

The general case maintains much of the character just described, even if the extreme inequalities between masses do not obtain. The results given here, together with those for the intermediate situation $m_1 = m_2$, give the range of possible behaviors of this type of system.

14.2   Small Oscillations of Many-Body and of Continuous Systems; Classical Fields

Let us consider the situation in which a large number, n, of coupled masses on springs undergo small oscillation (Figure 14.2); we shall eventually be interested in taking the continuous limit. Suppose that all the masses and all the spring

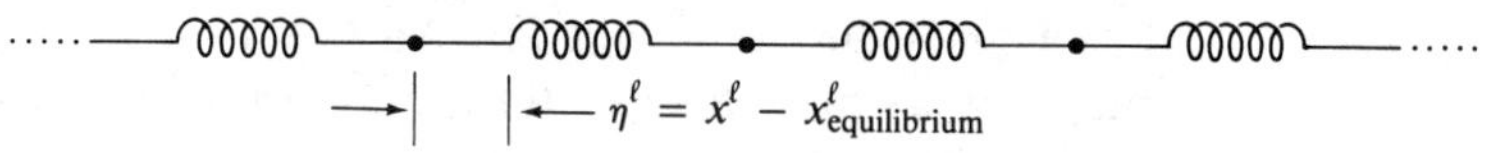

Figure 14.2

constants are the same. If we measure the position of the ℓth mass from its undisturbed position, we have, for the kinetic energy,

$$T = \tfrac{1}{2} \sum_{i=1}^{n} m(\dot{\eta}^i)^2. \tag{14.34}$$

The potential energy is

$$V = \tfrac{1}{2} \sum_{i=1}^{n+1} k(\eta^i - \eta^{i-1})^2, \tag{14.35}$$

where we impose the boundary conditions $\eta^0 = \eta^{n+1} = 0$.

We may proceed immediately to take the limit of a large number of in-finitesimal masses. The result is obtained by imagining the chain broken up into elements of mass dm. The equilibrium position of each dm is its "index" label. The variable $\eta(x)$ is now the separation between the element dm and its equilibrium position x. The kinetic energy goes simply over

$$T_{\text{cont}} = \frac{1}{2} \int_0^\ell \left(\frac{\partial \eta(x)}{\partial t} \right)^2 dm. \tag{14.36}$$

We can write $dm = \rho\, dx$, where ρ is a linear density. The potential energy is best handled by noting that the spring constant depends on the *length* of a spring, as well as on its more intrinsic properties such as composition, shape, and cross section. In fact, if two identical springs of constant k are hooked end to end, the resultant spring constant is $k/2$. This circumstance indicates that k is not an appropriate quantity in our limiting definition. Much more appropriate is the force per fractional extension $Y = \delta F/(\delta \ell/\ell)$. (Then Y is Young's modulus.) In Eq. (14.35) for the potential energy we note that $k(x^i - x^{i-1})$ is Y; hence

$$V = \frac{1}{2} \sum_{i=1}^{n+1} Y \frac{(\eta^i - \eta^{i-1})^2}{(x^i - x^{i-1})^2} (x^i - x^{i-1}). \tag{14.37}$$

In this form the continuous limit is immediate:

$$V_{\text{cont}} = \frac{1}{2} \int_0^\ell Y \left(\frac{\partial \eta}{\partial x} \right)^2 dx. \tag{14.38}$$

The Lagrangian for the motion of the string, $T - V$, yields the variation

$$\delta I = 0 = \int_{t_0}^{t_1} \int_0^\ell \left(\rho \frac{\partial \eta}{\partial t} \frac{\partial \delta \eta}{\partial t} - Y \frac{\partial \eta}{\partial x} \frac{\partial \delta \eta}{\partial x} \right) dx\, dt. \tag{14.39}$$

For the moment we assume that the ends of the string are fixed. Because the endpoints are fixed, the variations must vanish there, and the equation of motion for $\eta(x)$ becomes

$$\frac{\partial}{\partial t}\left(\rho\frac{\partial\eta}{\partial t}\right) - \frac{\partial}{\partial x}\left(Y\frac{\partial\eta}{\partial x}\right) = 0. \tag{14.40}$$

For constant ρ and Y this is a simple wave equation. In that case we see that a solution of the form $\eta = X(x)T(t)$ separates, giving

$$\frac{d^2T}{dt^2} = -\sigma^2 T, \tag{14.41}$$

$$\frac{Y}{\rho}\frac{d^2X}{dx^2} = -\sigma^2 X. \tag{14.42}$$

The solutions

$$\eta = e^{\pm i\sigma t}e^{\pm i\sigma x\sqrt{\rho/Y}} \tag{14.43}$$

(σ constant) are familiar as simple traveling waves if σ is real (but it need not be real). Notice that $\sqrt{Y/\rho}$ is the speed of the propagating disturbance.

A second way of handling the wave equation is to introduce coordinates along the **characteristics.** If $\sqrt{Y/\rho}\,(\partial/\partial x) \equiv \partial/\partial z$ and if

$$v = t + z, \tag{14.44a}$$

$$u = t - z, \tag{14.44b}$$

Eq. (14.30) reduces to

$$\frac{\partial}{\partial v}\frac{\partial}{\partial u}\eta = 0. \tag{14.45}$$

The immediate solution to this equation is

$$\eta = h\left(\sqrt{\frac{\rho}{Y}}\,x + t\right) + g\left(\sqrt{\frac{\rho}{Y}}\,x - t\right), \tag{14.46}$$

where h and g are completely arbitrary functions of their arguments. Note that the forms of solutions (14.43) and (14.46) are identical because any function of the form $h\left(\sqrt{\rho/Y}\,x + t\right)$, say, can be expressed via a Fourier integral of terms of the form $e^{i\sigma t + i\sigma\sqrt{\rho/Y}\,x}$. The forms (14.43) and the form (14.46) must be restricted by imposing the appropriate fixed-end boundary condition.

Let us return to the discrete system (rather than the limiting continuous system). The equation of motion can be obtained by a variation of $T - V$ where these energies are the discrete sums of Eqs. (14.34) and (14.35). Such a variation presents no difficulty if it is remembered that a particular coordinate η^s appears twice in the potential energy sum, in the combination $k(\eta^s - \eta^{s-1})^2 + k(\eta^{s+1} - \eta^s)^2$. We find the equations of motion:

$$m\ddot{\eta}^1 + k(2\eta^1 - \eta^2) = 0, \tag{14.47}$$

$$m\ddot{\eta}^i + k(2\eta^i - \eta^{i+1} - \eta^{i-1}) = 0 \qquad (1 < i < n), \qquad (14.48)$$
$$m\ddot{\eta}^n + k(2\eta^n - \eta^{n-1}) = 0, \qquad\qquad (14.49)$$

where we have singled out of the equation of motion the first and the last (nth) mass because they are distinguished by the fact that $\eta^0 = \eta^{n+1} = 0$. That is, here we are explicitly imposing the boundary condition. This system fits into the general scheme with $T_{ij} = m\delta_{ij}$ and

$$V_{ij} = k \begin{bmatrix} 2 & -1 \\ -1 & 2 & -1 \\ & -1 & 2 & -1 \\ & & -1 & 2 & -1 \\ & & & & \ddots \\ & & & & & 2 & -1 \\ & & & & & -1 & 2 \end{bmatrix}. \qquad (14.50)$$

To find the normal modes of this system, we must evaluate

$$\det(-\omega^2 T + V) = 0. \qquad (14.51)$$

To evaluate the determinant, we note that the determinant can be expanded out in cofactors of (say) the first row. Let D_n be the $n \otimes n$ determinant:

$$D_n = \det\left(-\frac{m\omega^2}{k}\delta_{ij} + k^{-1}V_{ij} \right). \qquad (14.52)$$

By expanding D_n in terms of its top row, we obtain two terms:

$$D_n = \left(2 - \frac{m\omega^2}{k} \right) D_{n-1} - D_{n-2}, \qquad (14.53)$$

where D_{n-1} and D_{n-2} are determinants of exactly the same form as D_n, but of $(n-1) \otimes (n-1)$ and $(n-2) \otimes (n-2)$ rank, respectively. (Hence they are the secular determinants appropriate to the discrete chain with $n-1$ and $n-2$ masses and the same boundary conditions as the n-mass case.) The simple tri-diagonal form of the **secular determinant** (14.50) allows us to write this recursion relation. Noting that $D_1 = 2 - m\omega^2/k$, while $D_2 = (2 - m\omega^2/k)^2 - 1$, we can use Eq. (14.53) to find the general D_n. A more economical approach is to note that Eq. (14.53) has a form reminiscent of a trigonometric identity. If

$$1 - \frac{m\omega^2}{2k} = \cos\theta \qquad (14.54)$$

is used to define an auxiliary variable θ, then Eq. (14.53) is identically the trigonometric equation,

$$D_n \equiv A\sin[(n+1)\theta] = 2\cos\theta A\sin(n\theta) - A\sin[(n-1)\theta], \qquad (14.55)$$

where A is an arbitrary constant, which can be fixed by noting that the quantity $D_1 = 2\cos\theta$ is, according to (14.55), equal to $A\sin 2\theta$. Hence $A =$

$2\cos\theta/\sin 2\theta$, and we have that

$$D_n = \frac{2\cos\theta}{\sin 2\theta}\sin\big[(n+1)\theta\big], \tag{14.56}$$

that is,

$$D_n = \frac{\sin\big((n+1)\theta\big)}{\sin\theta}. \tag{14.57}$$

The secular equation for the frequencies thus is

$$(n+1)\theta = N\pi, \qquad N = 1,2,3,\dots,n. \tag{14.58}$$

The value $N=0$ and the value $N=n+1$ are excluded because they lead to an indeterminate ratio $0/0$ or do not give zero for D_n in Eq. (14.57); besides, we expect exactly n different frequencies in the system. The frequencies are determined by Eq. (14.54):

$$\omega^2 = \frac{2k}{m}\left(1-\cos\frac{N\pi}{n+1}\right) $$
$$= \frac{4k}{m}\sin^2\frac{N\pi}{2(n+1)}. \tag{14.59}$$

The motion of the masses is calculated by means of Eqs. (14.47)–(14.49); writing $\eta^i = B^i e^{i\omega t}$ gives

$$(-m\omega^2 + 2k)B^1 = kB^2, \tag{14.60a}$$

$$(-m\omega^2 + 2k)B^i - kB^{i-1} = kB^{i+1}; \qquad i = 2,\dots,n-1, \tag{14.60b}$$

$$(-m\omega^2 + 2k)B^n - kB^{n-1} = 0. \tag{14.60c}$$

Recalling the definition of θ [Eq. (14.54)] yields

$$B^2 = 2\cos\theta B^1. \tag{14.61}$$

Suppose that $B^1 = \sin\theta$. Then $B^2 = \sin 2\theta$, and (14.60b) reads

$$B^{j+1} = 2\cos\theta B^j - B^{j-1}$$
$$= \sin[(j+1)\theta]; \qquad j = 2,\dots,n-1$$
$$= \sin\frac{(j+1)N\pi}{n+1}. \tag{14.62}$$

Notice that the last of Eq. (14.60) is identically solved because one could write

$$B^{n+1} = \sin\frac{(n+1)N\pi}{n+1}$$

on the right; but this B^{n+1} vanishes identically. Hence the boundary conditions are taken care of automatically.

Consider $N=1$. Then the B^j have all the same sign, increase smoothly as j increases, reach a peak for $j = \big[(n+1)/2\big]$, where the brackets denote the largest integer contained in $(n+1)/2$, and B^i then decrease smoothly to a small value for $j = n$. A plot of the amplitudes of B^j is just a half-wave of a sine function, and $N = 1$ clearly corresponds to the lowest mode of a string with fixed ends. The frequency of this mode is

$$\omega_1 \approx 2\sqrt{\frac{k}{m}} \frac{\pi}{2(n+1)}. \tag{14.63}$$

To compare this to a uniform string, recall that each individual mass relates to a linear density:

$$m = \frac{\rho\ell}{n+1}, \tag{14.64}$$

where ℓ is the total length of the string of masses.

Similarly, the Young's modulus is

$$Y = kd, \tag{14.65}$$

where d is the resting separation between masses: $d = \ell/(n+1)$. Hence

$$\omega_1 \simeq \sqrt{\frac{Y}{\rho}} \frac{\pi}{\ell}, \tag{14.66}$$

which, to the approximation $\sin\alpha \simeq \alpha$, $A \ll 1$, and $\eta \gg 1$, is the frequency of the lowest mode on a continuous string. Now take $N = 2$. To the extent that the small-angle approximation still holds, the frequency of this mode is $\omega_2 \simeq 2\omega_1$, just the relationship expected for the second mode on the stretched string. From Eq. (14.62) we see that the amplitudes of the individual mass motion vary as one full cycle of a sine wave, just as would be expected of the second mode of the string.

For modes that have larger N, the linear approximation for the frequency is no longer valid. In contrast to the situation for the continuous string, the discrete case has a highest frequency, ω_n:

$$\omega_n = 2\sqrt{\frac{k}{m}} \sin\frac{n\pi}{2(n+1)}$$

$$= 2\sqrt{\frac{Y}{\rho}\frac{n+1}{\ell}} \sin\frac{n\pi}{2(n+1)}. \tag{14.67}$$

Since if $n \gg 1$ the argument of the sin is $\sim \pi/2$, we see that the frequency approximates what would be expected if each mass were oscillating while attached to a spring of spring constant $4k$. The corresponding motion

$$B^j = \sin\frac{jn\pi}{n+1}, \tag{14.68}$$

shows that alternate mass points move in opposite directions. Hence each of the two springs attached to a mass is stretched to approximately twice the displacement of an individual particle, with the result $\omega^2 \sim 4k/m$.

14.3   Periodic Boundary Conditions

To model the bulk behavior of physical systems more correctly, it is often appropriate to consider periodic boundary conditions. An example of such a situation is provided in our previous discrete-mass model if we identify the $(n + 1)$st mass with the first. This has the effect of eliminating any preferred "edges" from our system. The equations of motion are like Eqs. (14.47)–(14.49) except that the equation for the first and for the nth masses have exactly the same form as those for all the other masses:

$$\ddot{\eta}^1 + k(2\eta^1 - \eta^0 - \eta^2) = 0 \qquad (14.69)$$

and

$$\ddot{\eta}^n + k(2\eta^n - \eta^{n-1} - \eta^{n+1}) = 0, \qquad (14.70)$$

where $\eta^0 \equiv \eta^n$ and $\eta^{n+1} \equiv \eta^1$. The secular determinant in this case is

$$\begin{vmatrix} -m\omega^2 + 2k & -k & & & -k \\ -k & -m\omega^2 + 2k & -k & & \\ & & \ddots & & \\ & -k & & -m\omega^2 + 2k & -k \\ -k & & & -k & -m\omega^2 + 2k \end{vmatrix} ; \qquad (14.71)$$

that is, the off-diagonal corners carry an additional entry $-k$ compared to our fixed-end example (a prescription that holds even in the case $n = 2$, in which case the corner entries are $-2k$).

Rather than directly solving the secular determinant, we return to the equations of motion and look for an x-dependence that is a sinusoidal function of the index of the point masses. Suppose that $\eta^j = B^j e^{i\omega t} = e^{iAj} e^{i\omega t}$. Periodicity requires $\eta^{j+n} = \eta^j$, which implies that $\exp(inA) = 1$; $nA = 2N\pi$. Equation (14.49) becomes

$$\left[-m\omega^2 + k(2 - e^{iA} - e^{-iA}) \right] e^{i\omega t} e^{iAj} = 0. \qquad (14.72)$$

Clearly, we have

$$\omega^2 = 2\frac{k}{m} (1 - \cos A)$$

$$= \frac{4k}{m} \sin^2 \frac{N\pi}{n}; \qquad N = 0, 1, 2, \ldots, n - 1. \qquad (14.73)$$

(The $N \equiv 0$ mode is a mode of uniform translation with $\omega = 0$.)

From the form of the solution, we have

$$B^j \propto e^{i2N\pi j/n}. \qquad (14.74)$$

The difference between the fixed boundary conditions and the periodic boundary conditions is the replacement of $2/n$ for $1/(n + 1)$ [cf. (14.73) with (14.59)]. For large n the two cases are strikingly similar.

For the periodic boundary conditions, the lowest nonzero frequency is

$$\omega \approx 2\sqrt{\frac{k}{m}}\,\frac{\pi}{n}$$

$$\approx \sqrt{\frac{Y}{\rho}}\,\frac{2\pi}{\ell}, \tag{14.75}$$

which corresponds to the second mode of a fixed-end string but is the lowest mode allowed by the periodic boundary conditions. The highest frequency is

$$\omega^2 = \frac{4k}{m}, \tag{14.76}$$

for which case

$$B^j = e^{i\pi j}, \tag{14.77}$$

showing that the amplitude is in opposite directions at adjacent mass parts, exactly as in the highest-frequency modes in the fixed-end case. Note also that the frequency reaches a maximum and then declines again, as the index N increases. (In the fixed-end case the frequency is maximum for $N = n$.) Let us introduce the reduced wave number, $k̸ = 1/\lambda$. Then a wave of spatial wavelength λ has the form $\exp(i2\pi \, k̸x)$. (Above we used $A \equiv 2\pi \, k̸\ell/n$.) For the periodic chain of masses we then have

$$\omega^2 = \frac{4k}{m}\,\sin^2\frac{\pi \, k̸\ell}{n} \tag{14.78}$$

where because of the discreteness of A, we have

$$k̸ = \frac{N}{\ell}, \qquad N = 0, 1, \ldots, n-1. \tag{14.79}$$

For small N, we have

$$\omega \approx \sqrt{\frac{Y}{\rho}}\,2\pi \, k̸, \tag{14.80}$$

which is the characteristic relation between wavelength and frequency on a stretched string. For shorter wavelength the relation $\omega(k)$ deviates from a straight line. The consequence is that short-wavelength perturbations oscillate with a lower frequency than that obtained by extrapolating (14.80). This is the phenomenon of **dispersion**; it means, for instance, that high-frequency waves propagate more slowly than low-frequency waves. Hence a sharp pulse will be spread out, or dispersed.

EXERCISES

14.1. Consider the motion of the following system. A square plate, side ℓ, thickness t, mass m, is supported by four springs of equal spring constant k. The plate is constrained so that it cannot rotate in its own plane;

that is, looking down on the plate, rotations in the plane about the center
are prohibited.

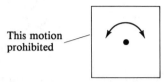

Figure P14.1

(a) Linearize the system about its equilibrium position (all springs of
equal length). What coordinates are you going to use, and how
many degrees of freedom does the system have?
(b) Set up the equations that you must solve to find the eigenfrequencies.
(c) Can you get the eigenfrequencies by a physical argument or special choice of coordinates? What are the eigenfrequencies and the
corresponding normal modes?

14.2. A heavy ring of wire (mass M) is suspended so that it always lies in
the same vertical plane; it hangs in the earth's gravitational field at the

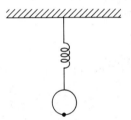

Figure P14.2

end of a spring and can move only vertically. A bead of mass m slides
without friction on the wire so that, in equilibrium, it rests at the lowest
point. Investigate the vibrations of this system, finding the normal modes
(eigenmodes) and the normal frequencies (eigenfrequencies).

14.3. Two equal masses m are connected to each other by a spring of spring
constant k, and in addition one of the masses is connected to a fixed
support by another spring of constant k. Motion is allowed only in a
straight line away from or toward the support.

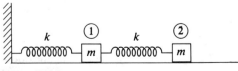

Figure P14.3

(a) Write the Lagrangian for this system. Obtain the equations of motion and obtain the frequencies and associated motions for the normal modes.

(b) Write out the explicit solution for the situation for which the initial condition has mass 1 at its equilibrium position but mass 2 positioned so that it is a distance A_0 from its equilibrium position (so that its spring is stretched).

14.4. Consider the double pendulum shown in the figure.

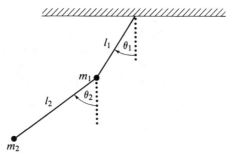

Figure P14.4

(a) Write down the exact Lagrangian for the system.

(b) Linearize the system (i.e., consider small oscillations about the equilibrium position $\theta_1 = \theta_2 = 0$).

(c) Solve the small perturbation problem: Find the eigenfrequencies of the motion.

(d) What are the normal modes of the motion? Describe the motion of the pendulum to which they correspond.

14.5. Obtain the small oscillation frequencies and modes for a block, mass M,

Figure P14.5

sliding on a smooth table and supporting a pendulum of length ℓ with a mass m at the end.

14.6. Consider the small oscillations about equilibrium of the following system. A hollow (massive) sphere is constrained to move only vertically and is supported by a spring of force constant k. Mass of sphere $\equiv M$. A small particle mass m is free to move inside the sphere, the inner surface of which is frictionless. At rest the small particle is at equilibrium at the lowest point in the sphere. Note that the sphere cannot rotate and can move only vertically!

   (a) How many degrees of freedom are there?

   (b) Write down the exact Lagrangian for the system, assuming that the particle remains in contact with the inner surface of the sphere.

   (c) Linearize the system and find the normal modes. Describe the motion represented by each normal mode.

14.7. Consider the following situation: A string of mass density μ hangs in a gravitational field, so that the tension T in the string is provided by its own weight. Analyze the traverse oscillations of the string-mass system. This is best done by using a variational principle.

INDEX

Note: **Boldfaced** words in the text are indexed, and all indexed words are in **boldface.** Emphasized words in the text are in *italic* and are not indexed.

Action, 39, 181
  action-angle formulation, Ch 12
  action variable, 189, 195, §12.2
Actual path, 40, 129
Adiabatic invariance/invariants, 197, §12.2, 208
Affine parameter, 44
Allowed path, 129
Angular momentum, 100
Antisymmetric (tensor), 17
Atlas, 5

Beam density, 59
Body frame, 104, 107
Boundary conditions, periodic, §14.3

Canonical: momentum, 42
  perturbations/perturbation theory, Ch 13, §13.1
  transformations, 129, Ch 10, 163

transformations, active infinitesimal, Ch 11, 175
  with respect to H, 164
Center of mass, 49
  CM frame, 63
Central force fields, Ch 4
Centrifugal force, 110
Characteristics, 234
Classical fields, §14.2
Closed form, 160
Commutator (of vector fields), 26
Compact (manifolds), 22
Completely separable, 188
Components: (of a tensor), 15
  (of a vector), 7
Configuration space, §1.4, 17
Conjugate (momentum), 42
Connected (manifolds), 22
Constant of the motion, 42

Constraints, holonomic, §1.5, 19
Continuous systems, Ch 14, §14.2
Contraction, 14
Contravariant vector, 13
Coordinates: coordinate patch, 4
  transformations/transformation law, 5, 15
Coriolis force, 110
Cosmological model, 45
Coulomb-force, §4.4
Covariant vector/covector, 13
Cross section, 59, 60
  differential, 60
  total, 62
Curl **d**, 150
  curl-free (form), 160
Curve, 6
Cyclic (variable), 42

d curl, 150
Determinant, secular, 235
Differentiable manifold, 5
Differential form: *see* Forms
Dispersion, 239
Dual (bases), 13
Dummy indices, 8

Eigenfrequencies, 230
Eigenvalues, 101
Electromagnetic field/field tensor, 111, 141, 159
Equivalent (Lagrangian), Ch 5, 76
  free particle, §5.1
  many dimensions, §5.3
  one-dimensional, §5.2
Euler: angles, 90
  equations (for rotating bodies), 105
  Euler's theorem, 123
Euler-Lagrange equations, 40
Exact (form), 160
Extended Hamiltonian, 132

Forms (differential form), 13, §9.2
  0-form, 150
  1-form, §1.2, 12, 150
  2-form, 152
  r-forms, 17, 150, 156
Fouling function, 80
Free indices, 8

Functional differentiation, 41
Fundamental theorem of calculus, 151

Gauge transformation, 161
Generating function/generator, 165
  infinitesimal, 96, 174
Geodesic, 44
  null, 44
Gradients, §1.2, 12
Grassmann algebra, 161

Hamilton's equations, §8.1, 119
Hamilton-Jacobi theory, §11.1, 175
  and wave mechanics, §11.3
  equation, 176
  time-independent case, §11.2
Hamiltonian/Hamiltonian systems, Ch 8, 118
  extended, 132
  flow, 175
  separable, Ch 12
Harmonic oscillators, coupled, §14.1
Holonomic constraints, §1.5, 19
Horizon (in cosmology), 45

Impact parameter, 59
Implicit function theorem, 21
Integrability theorem for 1-forms, 152
Integration, §9.2

Jacobi identity, 26, 148
Jacobian (determinant), 154
JWKB approximation, 184

Kepler problem, §12.3
  Kepler's second law, 55
  Kepler's third law, 71
Kinetic energy, 16, §1.4, 18
Kronecker delta, 11

Lab frame, 63
Lagrange: (Euler-Lagrange) equations, 40
  multiplier, 42
Lagrangians/Lagrangian systems, Ch 3
Larmor's theorem, §7.3, 113
Laurent series, 192
Legendre transformation, 118
Levi-Civita tensor, 52

Liouville's theorem, 175
Long-range (forces), 62
Lorentz force, 111, 141

Manifolds, §1.1, 4
Many-body systems, §14.2
Metric tensor, 44
Multiply: degenerate (systems), 207
    periodic (systems), 189

Noether's theorem, §8.2, 126, 127
Normal modes, 230

Orbit/orbit equation, 53, §4.3
Orientable (manifolds), 161
Orthogonal (matrices), 88
Oscillations, small, Ch 14

Partially degenerate (systems), 207
Path, 6
    actual, 40, 129
    allowed, 129
    potential, 40
Pauli spin matrices, 94
Perturbation, canonical, §13.1
Phase space/spacetime, 122, §8.3
Poincare invariants, 170
Poisson bracket, Ch 9, §9.1, 148
Potential path, 40
Proper time, 6
Pseudovector, 99

Range convention, 8
Rank (of tensor), 15
Residue (theorem), 192
Rigid body, Ch 7, §7.1
Rotations, Ch 6, 87
    group: algebra (of generators), 96
    proper, 91
    rotating frames, §7.2, §7.3

Rutherford cross section, 61

s-equivalent (Lagrangians), 76
Scattering, §4.5
    angle, 60
Schrödinger equation, 183
Secular determinant, 235
Separating constants (of the motion), 179
Simply connected (manifolds), 22
    separable, 188
Space frame, 107
Spacetime, Ch 2
Spinors, Ch 6, 91
Summation convention, 8
Symmetric (tensors), 17
Symplectic 2-form, 128-29, 167

Tangent: bundle, 9, 18
    space, 9
    vector, 6
Tensors/tensor fields, §1.3, 15
    r-contravariant, p-covariant, 15
    tensor product, 14
Three-body problem, §4.6
Trajectory, 53
Transformation, active ,94
Translation group, 97
Two-body problem, §4.2

Unitary (matrices), 92

Vectors/vector fields, §1.1, 6, 9
    differential operator, 7
Velocity, 6

Wave mechanics, Hamilton-Jacobi theory and,
    §11.3
Wedge (symbol), 153, 156
Weiss action principle, §8.2, 126